Sense and Nonsense

the writers of March   Portuguese
translator-chanteuse?

coffee - "gateway drug"

Lamarck

Darwin   Huxley   A.R.Wallace

Galton

Haeckel

Spencer

Douglas Spalding   (→ B. Russell)

Geo. Stanley Hall

Sense and Nonsense

Evolutionary Perspectives on Human Behaviour

Kevin N. Laland

Royal Society University Research Fellow
Sub-Department of Animal Behaviour
University of Cambridge

and

Gillian R. Brown

Research Scientist
Sub-Department of Animal Behaviour
University of Cambridge

OXFORD
UNIVERSITY PRESS

OXFORD
UNIVERSITY PRESS

Great Clarendon Street, Oxford OX2 6DP

Oxford University Press is a department of the University of Oxford.
It furthers the University's objective of excellence in research, scholarship,
and education by publishing worldwide in

Oxford New York

Auckland Bangkok Buenos Aires Cape Town Chennai
Dar es Salaam Delhi Hong Kong Istanbul Karachi
Kolkata Kuala Lumpur Madrid Melbourne Mexico City
Mumbai Nairobi São Paulo Shanghai Taipei Tokyo Toronto

Oxford is a registered trade mark of Oxford University Press
in the UK and in certain other countries

Published in the United States
By Oxford University Press Inc., New York

A catalogue record for this title is available from the British Library

Library of Congress Cataloguing in Publication Data
Data available

ISBN 0 19 850884 0 (Hbk)
10 9 8 7 6 5 4 3 2

Typeset by EXPO Holdings, Malaysia
Printed in Great Britain
on acid-free paper by T.J. International, Padstow, Cornwall

Preface

Can evolutionary theory help us to understand human
behaviour and society? Many evolutionary biologists,
anthropologists, and psychologists are optimistic that evolutionary principles can be applied to human behaviour,
and have offered evolutionary explanations for a wide range
of human characteristics, such as homicide, religion, and
sex differences in behaviour. Others are sceptical of these
interpretations, and stress the effects of learning and culture. They maintain that human beings are too special to
study as if they were just another animal—after all, we have
complex culture, language, and writing, and we build
houses and programme computers. Perhaps both of these
stances are right to a degree. Some aspects of our behaviour
may be more usefully investigated using the methods of
evolutionary biology than others. The challenge for scientists will be to determine which facets of humanity are
open to this kind of analysis, and to devise definitive tests of
any hypotheses concerning our evolutionary legacy. For
those of us fascinated by this challenge, knowledge of
the diverse methods by which human behaviour is studied
from an evolutionary perspective would seem a pre-
requisite. In this book, we outline five evolutionary
approaches that have been used to investigate human
behaviour and characterize their methodologies and
assumptions. These approaches are sociobiology, human
behavioural ecology, evolutionary psychology, memetics,
and gene–culture coevolution. For each, we discuss their

positive features and their limitations and in the final chapter we compare their relative merits.

Innumerable popular books have already been published that discuss human behaviour and evolution, e.g. *The Selfish Gene* (Dawkins, 1976), *The Rise and Fall of the Third Chimpanzee* (Diamond, 1991), *Darwin's Dangerous Idea* (Dennett, 1995), *How the Mind Works* (Pinker, 1997) and *The Meme Machine* (Blackmore, 1999). Each gives a unique and stimulating view of human nature. However, such books usually take a single viewpoint on human evolution, frequently identifying with a particular school, such as evolutionary psychology or memetics. There have also been academic books published from these different perspectives, such as *Culture and the Evolutionary Process* (Boyd and Richerson, 1985), *The Adapted Mind* (Barkow *et al.*, 1992), *Adaptation and Human Behavior* (Cronk *et al.*, 2000) and *Darwinizing Culture: the Status of Memetics as a Science* (Aunger, 2000). In contrast to these, our book takes a pluralistic approach, highlighting how different researchers have divergent views on the best way to use evolutionary theory to study humanity. Heated debates and personal attacks have often ensued. Some of the approaches described will be new to many readers, as the theories on which they are based have generally not made it further than the specialist scientific literature. In presenting these fields we endeavour to translate these methodologies into easily understandable examples, and thereby make accessible new perspectives on how human behaviour and culture can be interpreted.

In writing this book, we pursue three goals. First, like Eric Alden Smith and colleagues (Smith *et al.*, 2001), we see a need for 'a guide for the perplexed' for those of us who have struggled to understand the plethora of confusing terms and apparent differences of opinion and approach in the use of evolutionary theory to study human behaviour. Secondly, in line with a long tradition of researchers based at the Sub-Department of Animal Behaviour at the University of Cambridge, where we work, we believe that research in this domain is best served by a rigorous, self-critical science, and that the study of behaviour requires a broad perspective that incorporates questions such as how behaviour develops over an individual's lifetime as well as questions about how behaviour evolves. Thirdly, we see great value in pluralism in the use of methodology, and the integration of approaches. We hope to have made a small contribution in each of these regards.

This book does not provide an overview of the use of evolutionary theory in areas such as economics, law, and literature. We acknowledge the important work in these areas, but would rather maintain the length of the book as it is, and remain within more familiar territory. To those whose research is addressed, we hope that a fair synopsis is provided and are very grateful to all of the experts who have taken the time to discuss their work with us. We have personal views on the relative merits of the five schools of thought described; however, we have attempted to treat each approach evenly by asking leading members in the fields to help us to present their views accurately. Perhaps our profiles of the alternative approaches will highlight to some researchers how the methods may be integrated in the future, as well as draw attention to the conflicts that are yet to be resolved. Of those who currently deny the relevance of biology to the study of human behaviour, we hope that we might perhaps make some converts. More realistically, we hope that their scepticism will be tempered by the realization that not all researchers in this area are genetic determinists, Panglossian adaptationists, or wanton biologizers, and that many are prepared to place emphasis on non-biological and even non-evolutionary explanations.

Our intention is that this introductory book will be of use to undergraduate and postgraduate students (for example, in zoology, anthropology, and psychology) and to experts on one approach who would like to know more about the other perspectives, but also to lay persons interested in evolutionary explanations of human behaviour. We have tried to write the text so that anyone interested in this subject area will find the material easy to comprehend. Our intention is not to provide a textbook review of the whole subject area, but rather to give a taste of the various options. For readers who would like to know more about a particular perspective, further reading is provided at the end of the book.

The most enjoyable aspect of writing this book has been the opportunity to interact with many of the leading authorities in this area of research. We have been overwhelmed by the kindness and generosity of those who have discussed their work with us and have commented on chapters of the book: we have learned so much from them. We would like to thank the following people for commenting on one or more chapters and for discussing the material in the book: Robert Aunger, Pat Bateson, Gillian Bentley, Susan Blackmore, Monique Borgerhoff Mulder, Robert Boyd, Nicky Clayton, Tim Clutton-Brock, Leda Cosmides, Alan Costall, Nick Davies, Richard Dawkins, Daniel Dennett, Robin Dunbar, Dominic Dwyer, Marc Feldman, Dan Fessler, Jeff Galef, Oliver Goodenough, Russell Gray,

Kristen Hawkes, Robert Hinde, Sarah Blaffer Hrdy, David Hull, Rufus Johnstone, Mark Kirkpatrick, Richard Lewontin, Elizabeth Lloyd, John Maynard Smith, John Odling-Smee, Sally Otto, Henry Plotkin, Peter Richerson, Eric Alden Smith, Elliott Sober, John Tooby, Markus Vinzent, and Ed Wilson. We are also particularly grateful to Jeffrey Brown, Dominic Dwyer, Robert Hinde, Claire Laland, Bob Levin, Ed Morrison, and John Odling-Smee for reading the entire book and providing detailed feedback. We would like to thank the members of the Discussion Group at Madingley (Roz Almond, Yfke van Bergen, James Curley, Rachel Day, Tim Fawcett, Will Hoppitt, Jeremy Kendal, Bob Levin, and Liz Pimley), who worked through early drafts of each chapter with us, and provided very valuable input and encouragement. We were helped by comments from Mat Anderson, Martin Daly, Jean Dobel, Richard McElreath, Heather Proctor, and Joan Silk. Thanks also to Martin Baum at OUP and to Sheila Watson of Watson Little Ltd for their advice and guidance. This research was supported by a Royal Society University Research Fellowship to KNL and Medical Research Council funding to GRB. Finally, we are grateful to Ed Wilson and Sarah Blaffer Hrdy for their enthusiastic support and encouragement, the memories of which have kept us going when we thought that we might have bitten off more than we could chew.

K.N.L. and G.R.B.
March 2002

Contents

CHAPTER 1

Sense and nonsense

The human species is unique. We contemplate why we are here, and we seek to understand why we behave in the way that we do. Among the most compelling answers that modern science can provide for these eternal questions are those based on evolutionary theory. Few ideas have excited more reflection than Darwin's theory of evolution by natural selection. Currently, evolutionary thinking is everywhere. Up-and-coming young executives look to evolutionary lore for the latest in business acumen. Prisons use evolutionary logic to reduce tension among inmates. Medics exploit knowledge of human evolution to revise diagnoses and develop new treatments. Even grocery stores are taking on evolutionarily minded psychologists as consultants to tell them how best to stack their shelves.

Judging by its media profile and its representation in academic and popular science, evolutionary theory would seem to provide the solution to almost every puzzle. Every day, the newspapers abound with evolutionary explanations for human characteristics such as 'aggression' or 'criminal behaviour', while book shops are overflowing with popular science texts boldly asserting that evolution will reveal how to find your perfect partner, how to have a successful marriage, or how to make it to the top of your profession. We are told by various authors that our minds are fashioned to reason like hunter–gatherers, that we

With a uniqueness that signifies what?

behave like 'naked apes' floundering in a modern world, that rape is natural and male promiscuity inevitable, and that everything we do is ultimately a means to propagate our genes. However, to what extent can human behaviour be understood by taking an evolutionary viewpoint? What truth lies behind the newspaper reports and popular science stories? The aim of this book is to provide some answers to these questions.

Clearly, for many academic researchers, taking an evolutionary viewpoint is a fruitful means of interpreting human behaviour and society. Not only does evolution dominate the biological sciences, it increasingly makes inroads into the social sciences, with thriving new disciplines such as 'evolutionary psychology', 'evolutionary anthropology', and 'evolutionary economics'. Yet if an evolutionary perspective is so productive, why isn't everyone using it? What is it that leads the vast majority of professional academics in the social sciences not only to ignore evolutionary methods, but in many cases to be extremely hostile to the arguments? If evolutionary theory is having ramifications that permeate every aspect of human society, it would be reassuring to have confidence in the claims made in its name. In which case, should we not be concerned that some of the world's leading evolutionary biologists are highly critical of the manner in which fellow academics employ evolution to shed light on human nature?

The reality is that evolutionary perspectives on human behaviour frequently incite controversy, even amongst the scientists themselves. Evolutionary theory is one of the most fertile, wide-ranging, and inspiring of all scientific ideas. It offers a battery of methods and hypotheses that can be used to interpret human behaviour. However, the legitimacy of this exercise is at the centre of a heated controversy that has raged for over a century. Ultimately, the disquiet traces back to past misuses of evolutionary reasoning to bolster prejudiced ideas and ideologies. Although these transgressions often resulted from distortions of Darwinian thought, this darker side has resulted in many academic disciplines characterizing the use of evolution to elucidate humanity as harmful, even dangerous. Most researchers within the social sciences and humanities remain extremely uncomfortable with evolutionary approaches. Consequently, disputes over evolutionary interpretations of humanity have fostered a polarization of thought.

As evolutionary theory becomes more technical, many people find it difficult to distinguish basic biological truths from speculative stories or prejudicial argument. Like all areas of science, the work in this field varies greatly in quality. At its best, evolutionary analyses of human behaviour meet the highest standards, but at the other extreme we find a sensationalistic 'tabloid' pseudoscience. Zealous evolutionary advocates rarely admit to the difficulties that beset some of their more contentious revelations, while impassioned critics seldom acknowledge that there is some merit to an evolutionary analysis.

This book outlines the most prominent evolutionary approaches and theories currently being used to study human behaviour, guiding the reader through the mire of confusing terminology, claim and counter-claim, and polemic statements. We will explore to what extent human behaviour can legitimately be studied using these evolutionary methods. At the same time we will consider whether there are unique features of human society and culture that sometimes render such methods impotent. Both evolutionary arguments and the allegations of the critics will be subjected to careful scrutiny. By the end of the

book the reader will feel better placed to assess the legitimacy of claims made about human behaviour under the
name of evolution.

Taking the middle ground

An example of the controversy that can surround the use of
evolution to interpret human behaviour is provided by the
extraordinary response to an academic textbook written
by Edward O. Wilson, an eminent Harvard University professor. In 1975 Wilson produced an encyclopaedic book on
animal behaviour entitled *Sociobiology: the New Synthesis.*
While under normal circumstances textbooks on animal
behaviour rarely become bestsellers or arouse much media
attention, Wilson's tome was different. In the final chapter
of the book Wilson described how the latest advances in the
study of animal behaviour, particularly the insights of biologists Robert Trivers and Bill Hamilton, might explain
many aspects of human behaviour. He provided biological
explanations for a broad array of controversial topics,
including the differences between the sexes, human aggression, religion, homosexuality, and xenophobia. He also predicted that it would not be long before the social sciences
were subsumed within the biological sciences. Wilson's
book provoked an uproar and launched what is now known
as the 'sociobiology debate', which raged throughout the
1970s and 1980s. Social scientists bitterly disputed Wilson's
claims, found fault with his methods, and dismissed his
explanations as speculative stories. Intriguingly, among the
most prominent critics were two members of Wilson's own
department at Harvard, evolutionary biologists Richard
Lewontin and Stephen J. Gould, who vehemently attacked
the book in the popular press as simple-minded and reduc-

tionist. Yet most biologists could see the potential of the sociobiological viewpoint, which had paid great dividends in understanding other animals, and many were drawn into using these new tools to interpret humanity. The debate became polarized and highly political, with the sociobiologists accused of bolstering right-wing conservative values and the critics associated with Marxist ideology (more on this topic in Chapter 3).

In the midst of this controversy, when emotions were raised, and knee-jerk reactions common, the position of John Maynard Smith, one of the world's leading evolutionary biologists, stands out for its balanced judgement and fairness. In the heat of the debate, Maynard Smith retained a dignified intermediate position, supporting science over politics and being angry at much of the unjust criticism directed at Wilson, while at the same time remaining very conscious of the dangers of an inappropriate use of biology. In an interview in 1981, he stated:

> I have a lot of the gut feelings of my age of being horrified and scared of the application of biology to the social sciences—I can see...race theories, Nazism, anti-semitism and the whole of that. So that my initial gut reaction to Wilson's *Sociobiology* was one of considerable annoyance and distress (1981; quoted in Segerstråle, 2000, pp. 240–1).

Maynard Smith confessed to finding some of Wilson's views on human behaviour 'half-baked', even 'silly'. Yet in a balanced review of *Sociobiology* he described the book as making 'a major contribution' to an understanding of animal behaviour and was careful to stress its many positive features (Maynard Smith, 1975).

In her analysis of the sociobiology debate, sociologist Ullica Segerstråle (2000) states that few scientists were well positioned to be communicators or 'arbiters' between the sociobiologists and their critics, because few scientists understood both sides.[1] Indeed, opponents on either side of the debate had become so polarized and unreasonable that Maynard Smith later admitted that:

> I find that if I talk to Dick Lewontin or Steve Gould for an hour or two, I become a real sociobiologist, and if I talk to someone like Wilson or Trivers for an hour or two, I become wildly hostile to it (1981; quoted in Segerstråle, 2000, p. 241).

In this book, we endeavour to follow Maynard Smith's lead and take the middle ground between the positions of advocates of evolutionary approaches to the study of human behaviour and their critics. We hope that we have also provided a balanced, central view, which outlines the positive features of evolutionary methods but does not shy away from stating where we find the arguments suspect, and remains vigilant to the dangers of irresponsible biologizing. Some researchers appear to believe that all aspects of behaviour can be described by reference to human evolutionary history. We do not take this line, and believe that alternative explanations of human behaviour must be considered.

The high temperature of the sociobiology debate, and the severity of the criticism, would appear to have engendered a 'circle the wagons' mind set among human sociobiologists. When the flak was heavy they closed ranks, put up a united front, and some tacitly agreed not to criticize each other's work openly for fear of providing ammunition for the op-

---

[1] In addition to John Maynard Smith, Segerstråle (2000) singles out British ethologist Pat Bateson as an 'unusual scientist' who took the middle ground and played a mediating role between the protagonists.

position. At the founding meeting of the Human Behavior and Evolution Society (HBES) in Evanston in 1989, president Bill Hamilton gave an address in which he described scholars interested in the evolutionary basis of human behaviour as 'a small, besieged group' (Segerstråle, 2000). Some people present at the time recall Hamilton urging enthusiasts not to worry if their theories were crazy or their hypotheses untestable, but to march boldly ahead without fear of the consequences. One leading researcher, who was then a junior member of the society, recalls voicing the concern that this message would inadvertently foster a less rigorous approach to science, but this view received little support at the time. Other HBES members have told us that even today some resistance to self-criticism is apparent. We would not wish to stifle creativity which, after all, is one of the genuine benefits of an evolutionary perspective, and we recognize that there is a time for, and value to, brainstorming. Nonetheless, we believe that any scientific field needs to evaluate its own assumptions and research methods to progress, and that now that research into human behaviour and evolution is well established the strongest defence against external criticism would be to maintain the highest standards of science.

awk.

Within the broad community of researchers who take an evolutionary approach to investigate human behaviour, some individuals would appear to identify with particular subfields and see important distinctions between the approach of their subfield and that of the alternatives.[2]

---

[2] Those researchers who highlight the differences between approaches include Boyd and Richerson (1985), Symons (1989), Tooby and Cosmides (1989), Blackmore (1999), Hrdy (1999) and Smith *et al.* (2000). The counter-argument is put forward by Daly and Wilson (2000).

Others recognize no 'factions', and see no major differences in approach between the leading 'schools'. As the former position would appear to represent the views of the majority, in this book we characterize five different approaches to the study of human behaviour that have emerged since some key conceptual advances in the 1970s. These five approaches are human sociobiology, human behavioural ecology, evolutionary psychology, memetics, and gene–culture coevolution. As most researchers believe that the theory and methods of these subfields differ in important ways, we have emphasized these distinctions. Some of these differences may stem from their roots in different research traditions and academic disciplines while others are more ideological. In the final chapter of this book we compare evolutionary perspectives in an attempt to isolate which techniques are legitimate and insightful, and which are found wanting.

A guide for the bewildered

To the outsider, and even to many on the inside, the field of human behaviour and evolution is riddled with confusing terminology. There are 'Darwinian psychologists', 'evolutionary anthropologists', 'cultural selectionists', and 'gene–culture coevolutionists'. There are 'evolutionary psychology', 'dual-inheritance theory', 'human behavioural ecology', and 'memetics'. Some people cast all these approaches as 'human sociobiology' while others are at pains to distinguish between them. Until recently, Britain's most famous 'sociobiologist', Richard Dawkins, described himself as an 'ethologist', and was explicit about disliking the 'sociobiology' label.[3] In the Millennium edition of *Sociobiology: The New Synthesis*, Edward Wilson asserts that human sociobiology is 'nowadays also called evolutionary psychology' (Wilson, 2000, p. vii). However, Leda Cosmides and John Tooby, currently the world's most prominent evolutionary psychologists, deny that their discipline draws greatly from Wilson's sociobiology, while others disagree. When two other leading evolutionary psychologists, Martin Daly and Margo Wilson, published an article in which they described evolutionary psychology as 'the work of all those engaged in evolutionary analyses of human behaviour' (Daly and Wilson, 1999), they incurred the wrath of colleagues Eric Alden Smith, Monique Borgerhoff Mulder, and Kim Hill, who do not identify with this school (Smith *et al.*, 2000). Social scientist critics accuse evolutionists of ignoring cultural explanations of human behaviour, yet advocates of the 'meme' perspective provide an evolutionary explanation that is exclusively cultural.

One of our goals with this text is to lead the reader through this minefield of terms and concepts. In truth, there are many different ways of using evolutionary theory to study human behaviour and there is much disagreement within the field as to the best way to do it. This can result in confusion for outsiders, as well as for those who wish to use evolution themselves and are trying to distinguish between methodologies. What are the assumptions of each school?

---

[3] According to Segerstråle (2000) Dawkins described himself as an 'ethologist' in his books and writings up until 1985, when he finally classified himself as a 'sociobiologist' for strategic reasons. He wanted to counterattack on behalf of himself and others against the allegations in Rose, Lewontin and Kamin's (1984) *Not in Our Genes*.

Are some approaches more reliable than others? Are some right and others wrong? We discuss the history of using evolutionary approaches to describe human behaviour dating back to Darwin, which helps to explain why some of these divisions exist. Then, by comparing the different approaches, and critically evaluating their assumptions and methods, we hope to provide the information that the reader needs to assess which perspectives they will find the most compelling and which methods the most useful.

Asking evolutionary questions

The Nobel Prize winning ethologist Niko Tinbergen first suggested that there are four principal types of question that can be asked about a behaviour pattern (Tinbergen, 1963). Take an aspect of human maternal behaviour, for instance breast feeding. If one is investigating the behaviour of mothers to their babies, a researcher could ask: (1) What hormonal mechanisms and infant cues elicit breast feeding by the mother? (2) How does maternal care change over the lifetime of the mother as she becomes more experienced at raising children? (3) What is it about breast feeding that led to it being favoured by natural selection? Does it solely provide nutrition? Does it forge a mother–child bond? Does it confer protection against disease? (4) Why amongst humans do both parents care for their offspring when in other primate species parental care is largely restricted to mothers? The first question explores the *proximate* mechanisms or immediate causes underlying behaviour, while the second investigates the *development* of the behaviour during the lifetime of the individual. The third question addresses the *function* of the behaviour pattern and examines what advantage it gave our ancestors in the struggle to survive and reproduce. The fourth investigates the *evo-*

lutionary history of the behaviour and asks why a particular species is characterized by one trait rather than another. Questions of function and evolutionary history address different aspects of the evolution of a behaviour pattern.

In the book, we will see that different subfields place varying degrees of emphasis on the relative importance of these four classes of question. Disputes have arisen when protagonists have not clearly distinguished between these levels of analysis. We believe that answers are required on all of these dimensions to understand fully why a behaviour pattern occurs. One emphasis in the book will be that a full consideration of all four questions will provide the only complete description of human behaviour.

Another key issue to which we will repeatedly return is the value of making comparisons across species. Knowledge of how other animals behave can be of value in interpreting human behaviour. However, we must bear in mind that behaviour patterns that at first sight appear to be similar in human beings and other animals may in reality be entirely different. A good example is the male–male mounting behaviour observed in many monkeys, which has frequently been described as 'homosexual' behaviour (e.g. Bagemihl, 1999). There is, however, little evidence that male–male mounting in non-human primates and homosexuality in men share identical proximate causation, lifetime development, function, or evolutionary history: in non-human primates, same-sex mounting appears to play a role in social interactions and displays of dominance rather than providing a measure of sexual preference (Dixson, 1998). In this case, in spite of superficial similarities in activity, the causes of these behaviour patterns are almost certainly different for humans and other primates.

petitio principii

We can use another example to show what happens when evolutionary explanations are used to explain a trait before the relevant comparative evidence is well understood. Since the 1970s, scientists have asked, 'why do women have concealed ovulation?' Unlike the females of some other primate species, women exhibit no obvious sign that an egg has been released from an ovary and that they are approaching the time in the monthly cycle when sex is most likely to result in pregnancy. In fact, generally women don't know themselves on which day they ovulate. Female chimpanzees and baboons, on the other hand, advertise their time of ovulation with bright red swellings around their genitalia that are most fully swollen around midcycle when the female is most likely to conceive. When a female is fully swollen males will compete for the chance to mate with her, and females may copulate with several males during one ovarian cycle. In the light of these observations of closely related species, numerous evolutionary hypotheses have been proposed to explain what was it about our evolutionary past that led to selection for ovulation in women to be concealed—the function of concealed ovulation. For example, Alexander and Noonan (1979) suggested that concealment of ovulation would force a male to stay watching over the female throughout the full cycle, which would prevent him from seeking other partners. As a consequence, the man would be more certain that any offspring were his, while the woman would gain help from the father in looking after the children.[4] Other researchers went on to speculate that, if males were no longer competing over access to fertile females, the decreased tension within the group may have made cooperation between males (for example, dur-

---

[4] This idea has been further developed by Strassmann (1981) and Turke (1984).

ing hunting) more likely to evolve (Daniels, 1983). Nancy Burley (1979) put forward the alternative argument that women who had knowledge of their time of ovulation might actively avoid sexual intercourse around this time, in order to avoid the pain and risks of labour, and the costs of rearing a child. If this is the case women without any knowledge of their time of ovulation might leave greater numbers of descendants than women with this knowledge, leading to the selection of concealed ovulation.

We can therefore see a proliferation of ideas regarding the supposed evolutionary history and function of concealed ovulation. The problem with all of these hypotheses is that concealed ovulation is probably not a derived trait in human beings (Daly and Wilson, 1983; Burt, 1992; Pawlowski, 1999). In other words, it is not 'concealed ovulation' among our ancestors but 'advertised ovulation' in other species that evolved. Although common and pygmy chimpanzees have visible signs of ovulation, there is no reason to presume that the ancestors of chimpanzees and human beings had these swellings. As the majority of primate species, including most apes, do not reveal their time of ovulation, the possibility that chimpanzees evolved revealed ovulation after splitting from their common ancestors with human beings is more likely. If this is the case, the wrong question has been investigated. Rather than asking 'why do women have concealed ovulation', we should ask 'why have females of some primate species evolved obvious signals of ovulation?' Whether or not a particular trait has been subject to natural selection is one of the recurring problems that bedevil evolutionary analyses.

The study of human behaviour can derive much useful information from the behaviour of animals, particularly the other primates (Hrdy, 1999; Brown, 2000). Indeed, a comparative analysis is a critical step towards determining which evolutionary question to ask. However, this example also reveals how we must be wary of labelling a behaviour as an evolved trait without testing this assumption, and illustrates how evolutionary analyses may sometimes be mistaken.

Human culture, learning, and genetic determinism

The titles of popular science books taking an evolutionary perspective have described human beings variously as 'naked apes', 'scented apes', 'lopsided apes', or 'aquatic apes', and have referred to 'man the hunter' and 'mother nature'. Additionally, we have been told 'how the mind works', 'why sex is fun', and have had 'consciousness explained'. However, can there ever be a straightforward evolutionary explanation of human behaviour? Isn't there something different about human beings compared to our primate cousins and other animals? We have a complex culture, built around a spoken language and written texts. Surely human behaviour cannot be explained by our biology alone, as our culture sets us apart? For most social scientists human behaviour is largely learned from other people. Consequently, the principal reason why the people of New York differ in how they think and in what they do from the Ache hunter–gatherers of Paraguay or the Arctic Inuit of Canada is thought to be because they have been exposed to divergent cultures or had different social experiences. For social scientists, culture is most commonly regarded as a cohesive set of ideas, beliefs, and knowledge that exists in a completely different realm to biology. These researchers believe culture is the primary influence on human behaviour.

In contrast, many evolutionary-minded researchers think about culture more broadly as the product of an evolutionary process. In many animal species, individuals grow up in an environment that contains other individuals of the same species, and most primates exhibit complex societies (Smuts *et al.*, 1987). Moreover, many animals acquire skills and knowledge by learning from others, frequently adopting the 'cultural' traditions that characterize their population, often mediated by sophisticated forms of communication (Heyes and Galef, 1996). A recent scientific paper reported 39 distinct behaviour patterns maintained as cultural traditions in some populations of chimpanzees but not others, including distinct patterns of tool usage, courtship behaviour, and even medicinal skills, with each population's cultural repertoire handed down by one generation to the next (Whiten *et al.*, 1999). Of course, there are important differences between animal and human cultures, but there are likely to be some continuities between them too.

However, the five evolutionary approaches differ in the way in which they regard human culture, and the importance that they attribute to it. We shall see that some regard human culture as shaped by genetic biases and predispositions, and stress that there is much more uniformity to human behaviour and society than is given credence by traditional social scientists. They argue that there are hidden commonalities that are found universally across all societies; for instance, all cultures are structured by statuses and roles, and possess a division of labour (Brown, 1991). Others think of culture as the outcome of an interplay between our unusually flexible developmental systems and particular aspects of the ecological and social environment, an interplay that typically results in adaptive human behaviour. Perhaps seemingly arbitrary traditions for hunting particular animals or food preparation habits are actually the optimal solution to these problems given local conditions. Still others conceive of culture as an evolutionary process in its own right, with human minds adopting variant ideas in a similar manner to how genes are selected in biological evolution. Maybe scientific theories or political ideologies change over time in an equivalent manner to biological evolution. Finally, we shall come across a group of biologists and anthropologists that, like the majority of social scientists, see culture as socially transmitted information that passes between individuals, but focus on the interaction between genetic and cultural processes. For instance, perhaps we are predisposed to learn to be right-handed, but the frequency of right-handedness varies across cultures because of society-wide differences in their tolerance of left-handers.

The alternative evolutionary approaches also express quite different conceptions of the relationship between genes, development, learning, and culture. Some researchers regard developmental processes, including our capacity to learn for ourselves, as tightly constrained by a genetic straitjacket. From this viewpoint, we are programmed to learn that which in the evolutionary past enhanced our survival and reproduction, and society reflects these evolved imperatives. For instance, perhaps we are predisposed to acquire a fear of snakes or spiders because these creatures constituted very real dangers for our distant ancestors. Others regard development as much more flexible, and learning as only loosely guided by our genes, so that these processes can generate behavioural outcomes that are unspecified by prior evolution. For example, rather than evolving a specific dietary preference for fried fish or chocolate, maybe evolution has furnished us with a tendency to eat whatever happens to taste good, as our taste buds have evolved to detect foods with the energy and nutrients to promote health and well being. Differences in ideas about culture and learning will be highlighted in the later chapters.

One important point that needs to be made before we go any further is that using evolutionary theory is not the same as taking a genetic determinist viewpoint. Genetic determinism is the belief that our genes contain blueprints for our behaviour that will always be followed and that constitutes our destiny. Such a belief would run contrary to much that is known about how human behaviour develops. Where researchers talk about genetic influences on human behaviour, they do not mean that the behaviour is completely determined by genetic effects, that no other factors play a role in our development, or that a single gene is responsible for each behaviour. While most evolutionary biologists focus exclusively on genetic inheritance, it does not follow that they believe that genes are the sole determinant of human behaviour, and the vast majority take it for granted that multiple environmental influences will play a part throughout development. We will come across evolutionists that describe 'genes for' a particular trait (e.g. Dawkins, 1976), by which they mean genetic variation that, along with a multitude of environmental factors, affects a character. While this shorthand has been criticized as misleading by other biologists (e.g. Bateson, 1981), and while it may sometimes lead researchers to underestimate the importance of developmental processes, an evolutionary perspective does not equate with a genetic determinist view of human behaviour.[5]

---

[5] See Bateson and Martin (1999) for further discussion of genetic influences and behavioural development.

After decades of debate about the relative importance of 'nature' *versus* 'nurture', researchers have come to the rather uninspired conclusion that both nature (generally associated with genes) and nurture (typically representing environmental factors, learning, and culture) will obviously be of importance. So where do we go from here? Should biologists concentrate on determining how much of behaviour can be explained by genetic inheritance, while the social scientists are left alone to discuss human cultures and social structure? We think not. Most biologists have long rejected this dichotomous mode of reasoning. While we constantly hear reports in the press that scientists have detected 'the gene for' some trait such as breast cancer or schizophrenia, this language is highly misleading. Genetic and environmental influences on human behaviour are like the raw ingredients in a cake mix, with development analogous to baking (Bateson and Martin, 1999). As nobody expects to find all the separate ingredients represented as discrete, identifiable components of the cake, so nobody should expect to find a simple correspondence between a particular gene and particular aspects of an individual's behaviour or personality. Indeed, developmental biologists are agreed that the very idea that an individual's behaviour can be partitioned into nature and nurture components is non-sensical, as a multitude of interacting processes play a role in behavioural development (Bateson and Martin, 1999; Oyama *et al.*, 2001).[6] From this perspective, a complete

---

[6] Some researchers have attempted to partition the variance in human behavioural and personality traits, observed across many individuals in a population, into components that are due to genetic and environmental differences. Frequently such analyses are based on studies of identical and fraternal twins. However, such methods require complex statistical procedures, based on a number of key assumptions, and, as we shall see in Chapter 7, they are contentious.

understanding of human behaviour will only result from us studying human beings as animals developing in a rich social environment and immersed in complex cultural traditions.

Evolutionary perspectives on human behaviour

The history of using evolutionary ideas to interpret human behaviour is no dry and dreary chronicle of academic ideas. For a century and a half evolutionary thinking has had a dramatic influence on how human beings regard themselves, and in how societies structure their shared values, institutions, and laws. In Chapter 2 we provide an overview of these events. We begin with Charles Darwin, who wrote at great length about human beings. Darwin accumulated vast evidence that the gulf in mental ability between human beings and other animals was not as great as hitherto believed, showing both that animals are capable of surprisingly intelligent behaviour and that humans exhibit hidden brutish tendencies. We will also meet one of Darwin's relatives, Francis Galton, a brilliant scientist who devised the methods for using identical twins to investigate genetic influences on human behaviour. However, Galton was strongly biased towards biological explanations for human behaviour and mental abilities, which provided the basis for his writings on eugenics and founded a movement that years later was to result in discrimination and enforced sterilization. We shall see that Darwinian views on evolution were distorted into Social Darwinism, which applied a 'survival of the fittest' doctrine to social institutions, and used erroneous evolutionary arguments to argue that socialism was harmful and to justify unrestrained capitalism. We also see how evolutionary-minded anthropologists and biologists in the 19th century, confusing evolution with progress, applied the ideas of natural selection to the evolution of human societies and argued that some 'races' had reached a higher level of evolution than others. Darwinian ideas were to have a major influence on the theories of human development within psychology. For instance, Sigmund Freud took Darwin's ideas of sexual selection and the 'instinct' to mate and used them to develop his concept of the libido, a core of chiefly sexual urges that are the major underlying force behind human behaviour. We then move into the 20th century and discuss how evolutionary ideas influenced the conflict between ethologists and psychologists over the relative importance of instinct and learning. In the 1960s, popular ethology books, such as Konrad Lorenz's *On Aggression* and Desmond Morris's *The Naked Ape* were to introduce dubious and sensationalistic evolutionary arguments to the general public, and create major furores. While we also describe the many positive ramifications of evolutionary theories of humanity, this history helps us to understand why many people remain wary of applying evolutionary reasoning to humans, and helps us to understand the backgrounds from which modern approaches emerged.

In Chapters 3 to 7, we present five more recent evolutionary approaches to the study of human behaviour. Rather than providing a comprehensive overview of each subfield, we aim to give the reader a little taste of each of the alternatives. In all cases, we provide an introduction that shows how the subfield arose and which researchers played important roles in its development. This is followed by an account of the key ideas and methods that characterize the viewpoint, and a description of some of the more interesting pieces of research carried out by practitioners that illustrate the reasoning, merits, and findings of the particular school. Each chapter ends with a critical analysis of the beliefs and methods of the subfield in which we attempt an impartial evaluation of the arguments made and the tools used by those researchers, and discuss the main criticisms that have been levelled against each approach.

Contemporary evolutionary perspectives on human behaviour began in the 1960s and 1970s with a series of exciting breakthroughs in the study of animal behaviour that precipitated a revolution in evolutionary thought. Important new ideas such as *kin selection, reciprocal altruism,* and *evolutionary game theory* emerged through the work of Bill Hamilton, Robert Trivers, and John Maynard Smith, and these were to alter the course of zoology. In Chapter 3, we depict these novel theories and methods, which came together under the term 'sociobiology', and were brought to the attention of many through the books of Edward Wilson and Richard Dawkins. We describe how these ideas were applied to human behaviour and evaluate the political and scientific outcry that ensued. The principal charges made by the critics of human sociobiology are also examined, namely, that researchers had devised simplistic and prejudicial theories. We highlight important ideas that emerged from human sociobiology, such as the careful comparison of human behaviour with that of other animals, which can be seen to continue to this day to enlighten our views on human nature. While human sociobiologists were accused of abusing science to reinforce traditional values, we will give examples of sociobiological research that challenged stereotypes concerning human sex differences. Finally, we will describe how the field triggered the development of the four major contemporary approaches, human behavioural ecology, evolutionary psychology, memetics, and gene–culture coevolution. Almost certainly because of the controversy that surrounded it, few of today's researchers describe themselves as 'human socio-biologists', although there are notable exceptions.

In Chapter 4, we describe the field of human behavioural ecology that has continued to employ methods devised to study animal behaviour to ask questions about human beings. These investigators, many of whom have back-grounds in anthropology, are interested in exploring to what extent the differences in human behaviour can be explained as adaptive responses to the habitat in which they live. Human behavioural ecologists frequently construct mathematical models to compute the optimal human behaviour in a given context on the assumption that this is what might have evolved. They then test the model's pre-dictions, primarily studying traditional societies such as hunter–gatherers. We will see that these researchers claim to have found evidence that people choose food items in order to maximize their caloric returns and that they hunt in optimally sized groups. They assert that they are able to predict whether parents will have another baby given knowledge of the number of children parents already have and their wealth. Most extraordinary of all, they have devised evolutionary explanations for why parents in mod-ern, post-industrial societies may most effectively pass on their genes by having fewer children. But do people really behave in an adaptive or optimal manner? Critics suggest not, and declare that the research programme of human behavioural ecology is fundamentally misguided because it investigates the current function of behaviour instead of testing hypotheses concerning the evolved mental processes that guide behaviour. We will investigate to what extent these concerns are warranted.

In Chapter 5, we introduce the burgeoning new field of evolutionary psychology. These researchers are primarily academic psychologists interested in the evolved psychological mechanisms that underlie human behaviour, and who see modern human beings as creatures adapted to the environments of our stone-age ancestors. They use this idea to discuss how behaviour patterns that may have no apparent utility in our modern environment are more easily understood if we reconstruct how natural selection was acting in the past when our ancestors were hunter–gatherers. Evolutionary psychologists claim to have identified a number of mental adaptations which they believe regulate human behaviour even in modern societies, such as a tendency to be particularly sensitive to individuals that might be cheating on social rules, or for men to be more violent than women. Researchers report that across all continents there are universal sex differences in the characteristics that men and women look for in a partner, with men seeking to mate with many young women, and women choosing to devote themselves to a wealthy and powerful man. Evolutionary theory has been employed to provide explanations for such sex differences. However, this research programme has also attracted considerable criticism, as many observers fear that insufficient is known about our ancestors' way of life to be able to generate reliable hypotheses about the present (Rose and Rose, 2000).

In Chapter 6, we will evaluate the field of memetics, and investigate the hypothesis that culture exhibits its own evolutionary process. Oxford zoologist Richard Dawkins first introduced the concept of the 'meme' in his book *The Selfish Gene*, published in 1976. The main idea here is that aspects of our behaviour and knowledge, such as particular

skills, songs, ideas or rituals are transmitted between individuals through imitation and other forms of social learning. 'Meme' is the name given to such units of culture and,
as some memes are more likely to spread than others, there
is a new kind of evolution generated at the cultural level.
Somewhat disturbingly, the selection of one meme over
another may be of no advantage to the individual human
being; rather the meme makes use of us in order to replicate
itself. Memeticists suggest that human beings may behave
the way they do not because it is in their interests but
because their minds have been infected by a cultural virus.
Could consciousness be little more than a collection of
memes? Are the dominant world religions neither true nor
even beneficial, but merely those complexes of religious
ideas that happen to be best at spreading? Memetics has
been discussed at length in recent years, and has generated
many provocative hypotheses. However, it has spawned
little empirical work, and its critics describe memetics as
speculative evolutionary story-telling. At the end of this
chapter, we provide some ideas about how a useful and rigorous research programme for memetics could be devised.

In Chapter 7, we see that a quantitative science that shares
some similarities with memetics already existed, namely
gene–culture coevolution. However, these researchers
believe that biological and cultural evolution interact in
complex ways. Consequently, they use mathematical
models devised from population genetics theory to predict
how cultural traits spread through human populations by
social learning, and how genes and culture coevolve. For
these researchers, the last two million years is dominated by
this coevolution of genes and culture, which generates new
evolutionary mechanisms and transforms evolutionary
rates. The models show how cultural practices can have

important implications for genetic evolution. For instance, while most Western people can drink milk without getting sick, the majority of adult human beings cannot because they lack a gene partly responsible for the enzyme that breaks down lactose (Simoons, 1969; Durham, 1991). Intriguingly, those adults that can consume dairy products typically belong to cultures with a long tradition of dairy farming. Could the cultural practice of dairying have created the selection pressures that led some adult humans to be able to drink milk without becoming ill? Gene–culture models also provide new methods for partitioning the variance in human personality traits. We regularly hear reports that scientific studies using identical and non-identical twins have revealed a genetic explanation for differences between people in particular characteristics such as intelligence, but the gene–culture analyses challenge these findings from behavioural genetics. However, gene–culture coevolutionary methods are also subject to criticism. For instance, some social scientists have objected to the idea that culture can be analysed as if composed of discrete psychological or behavioural characteristics, while others have questioned the legitimacy of 'borrowing' biological models to account for culture. We will investigate whether cultural and genetic processes are too different for the former to be well described by models based on the latter.

In Chapter 8, all of these fields are brought together for comparison. Advocates of each approach often claim to have the foremost or the only valid perspective on evolution and human behaviour, and protagonists from different schools sometimes scrap amongst themselves. But which approach is best? Does each school exhibit strengths and weaknesses, or is one method superior to, or more legitimate than, the other? Could the different approaches be integrated into a single, overarching perspective that synthesizes techniques from disparate schools, or are there fundamental incompatibilites such that if one school is right another must be wrong? What exactly are the key differences of opinion, and how can they be resolved? After comparing the alternative views, and examining their ideological and methodological differences, in the final chapter of this book we assess to what extent it is possible to cross the boundaries between approaches and integrate them into a broad yet rigorous evolutionary science of human behaviour.

Sense and Nonsense endeavours to provide the reader with an informed account of alternative evolutionary perspectives in the hope that they will be better able to distinguish between them and to learn from them in a discerning manner. Having completed this book we hope that the reader will have acquired the necessary knowledge and skills to be able to evaluate evolutionary hypotheses concerning humanity for themselves, and to make their own judgements as to what makes sense and what is nonsense.

CHAPTER 2

A history of evolution and human behaviour

Few ideas have contributed as much to biological knowledge as Darwin's theory of evolution by natural selection, and yet this revolution within biology is just the tip of the Darwinian iceberg. 'Natural selection' has proved an irresistible abstraction, with countless scientists, social scientists, politicians, and business leaders drawn to its explanatory power. Not surprisingly, since publication of *The Origin of Species* in 1859, there has been a long history of using evolution to interpret human behaviour and society, some of which makes distinctly disturbing reading. As Maynard Smith (1975) pointed out:

> Attempts to import biological theories into sociology, from social Darwinism of the 19th century to the race theories of the 20th, have a justifiably bad reputation.

In this chapter, we trace the history of using evolutionary approaches to study human behaviour from the 1850s to the 1960s. We will see that evolutionary ideas were important in shaping our concept of human nature, sometimes bolstering racism and sexism while at other times dispelling unjust views. In fact, the last century and a half have been

derogatory ?

criticism ?

Find M. Sengels eugenicist writer.

characterized by constant battles between evolutionary advocates and their critics, frequently coinciding with a regular swinging of the pendulum to favour explanations for human behaviour in terms of nature or nurture. We illustrate how evolutionary arguments have been put forward as pretexts to justify the eugenics movement, Nazism, unfettered capitalism, racist immigration policy, and enforced sterilization, as well as to argue that some 'races' were more advanced than others. The vast majority of these assertions employed crude distortions of Darwin's theory, which derive more from the work of other 19th-century intellectuals such as Jean Lamarck and Herbert Spencer, although it is Darwin's name that is often unfairly linked to these views. We will also describe good works done by evolutionary biologists that counter racism and prejudice in society, and reveal countless important scientific insights and advances that followed from an evolutionary viewpoint. However, the abuses of evolutionary theory are more often remembered.

This historical perspective provides a context within which we can begin to interpret contemporary disputes over the use of evolution. For instance, it helps us to understand why the vast majority of social scientists are so resistant to evolutionary hypotheses. It also helps to explain why E. O. Wilson's *Sociobiology* was to provoke such profound hostility as to culminate in his physical attack by protestors (more in Chapter 3) and perhaps why many contemporary evolutionary psychologists place emphasis on the universal features of human nature (see Chapter 5). In the rest of the book, we will show how modern evolutionary approaches have increased our understanding of human behaviour, but we should remain aware of the social impacts that scientific theories can have.

Darwin's views on human behaviour

The history of using evolutionary theory to make sense of human behaviour begins with Charles Darwin. This is not only because Darwin was the first person to come up with a credible explanation for evolution, namely the process of natural selection, but also because Darwin wrote at great length about human beings.

In *The Origin of Species* (1859), Darwin patiently explained in a series of logical steps how natural selection works. Struck by the views of Thomas Malthus that population growth would eventually reach a point at which insufficient food was available, Darwin suggested that those individuals in the population whose anatomical, physiological, and behavioural characteristics best fitted the environment would have the greatest chances of surviving and reproducing. If those characteristics were heritable, then the next generation would contain a higher frequency of individuals with these 'fitter' traits, and hence the population would change over time. At the time of publication, the dominant view of the natural world was that each species had been individually created and was immutable. The ability of natural selection to explain how variation among individuals may lead to adaptation of species to their environments and the origin of new species has been confirmed by countless experiments and is now beyond dispute (Endler, 1986a; Jones, 1999).

The striking feature of *The Origin of Species* is that Darwin does not mention human evolution, except to say in the final pages that:

> In the distant future I see open fields for far more
> important researches. Psychology will be based on a new
> foundation, that of the necessary acquirement of each

mental power and capacity by gradation. Light will be
thrown on the origin of man and his history. (1859, p. 458)

An eager public had to wait over a decade for Darwin to
elaborate on these enigmatic statements. The idea that
human beings had evolved became the source of intense
public interest and hostility in the 1860s, but Darwin, fearing persecution and ridicule, refused to be drawn further
on human origins until a watertight case could be made
(Bonner and May, 1981). Instead, Darwin's great supporter
Thomas Huxley tenaciously fought his corner for him,
trouncing Bishop Wilberforce in a famous debate at Oxford
University in 1860. Huxley presented lectures and published *Evidence as to Man's Place in Nature* (1863), in which
he used the skeletons of apes to provide undeniable evidence that human beings were of animal ancestry. Among
archaeologists, the hunt for the remains of the 'missing link'
between humans and other apes had begun.

By 1870, Thomas Huxley, 'Darwin's Bulldog', had become
the prophet of the new world of science (Desmond, 1997).
Indeed, largely through Huxley's efforts, being a 'scientist'
became a legitimate profession, and 'science' came to exert
a major political influence, with Darwinism providing the
focus of this development (Desmond, 1997). By the 1870s,
Darwin was famous and everyone was waiting to hear what
the great man had to say on human evolution (Bonner and
May, 1981). With characteristic caution, Darwin eventually
brought forth *The Descent of Man and Selection in Relation
to Sex* (1871) and *The Expression of the Emotions in Man and
Animals* (1872), two huge monographs that were originally
intended to be a single work. Rather than dwelling on
human anatomy, Darwin drew attention to the question of
the evolution of mental ability, for which there seemed to
be a much greater divide between human beings and other

animals. He maintained that there was variation in mental capacity both within and between species, and suggested that being intellectually well endowed was advantageous in the struggle to survive and reproduce.

> To avoid enemies, or to attack them with success, to capture wild animals, and to invent and fashion weapons, requires the aid of the higher mental faculties, namely, observation, reason, invention, or imagination. (1871, p. 327)

Darwin sought to demonstrate that the differences in mental ability between human beings and other animals were not as great as widely believed. In contrast, Alfred Wallace, who had struck upon the idea of evolution by natural selection around the same time as Darwin, concluded that the complex language and the music, art, and morals of human beings could not be explained solely by natural selection and must have resulted from the intervention of a divine creator during human evolution (Wallace, 1869). Darwin attempted to counter the widespread belief that animals were merely machines driven by in-built mechanisms, while human beings alone were capable of reason and advanced mental processing. He attacked this dichotomy from both sides, arguing that human beings had more brutish tendencies, and animals more elevated intelligence, than hitherto conceived. In the first part of *The Descent of Man*, Darwin documented the evidence that human beings have a number of behavioural characteristics in common with other animals, including 'self-preservation, sexual love, the love of the mother for her new-born offspring, and the power possessed by the latter for suckling' (1871, p. 36). Similarly, in *The Expression of the Emotions*, Darwin catalogued an amazing array of equivalent facial expressions in humans and animals. By pointing out the striking similarities between the expressions associated with particular emotions in human beings and other animals, Darwin dismissed the theory that expressions had been uniquely given to human beings in order to communicate their emotional states to others. For instance, Darwin noted that apes and monkeys, like human beings, have 'an instinctive dread of serpents' and will respond to snakes with the same screams and the same fearful faces as many of us do. He described how one day he mischievously placed a stuffed snake into the monkey enclosures at London Zoo and the poor creatures 'dashed about their cages and uttered sharp signal-cries of danger, which were understood by the other monkeys' (1872, p. 43). Darwin also noted, around a century before modern researchers (Goodall, 1986), that chimpanzees use stone tools to crack open nuts, which suggested even less of a gap between the mental lives of human beings and apes than many Victorians in Britain wished to believe.

Darwin charmingly described the emotional lives of other animals in distinctly human terms. Even invertebrates were thought to feel pleasure and pain, happiness and misery, and show some intelligence. He maintained that, for all animals, 'terror acts in the same manner on them as on us, causing the muscles to tremble, the heart to palpitate, the sphincters to be relaxed, and the hair to stand on end' (p. 39). He described young ants chasing and pretending to bite each other, just like puppies, arguing that they are excited by the same emotions as us. He also maintained that courage and timidity are seen in dogs, that horses can be sulky, and monkeys vengeful.

Judged by contemporary standards, these arguments are naive, anthropomorphic, and anecdotal. Yet many of Darwin's most fundamental assertions about animal mental abilities have been proven correct. Most researchers into animal behaviour would agree that many animals do feel pleasure and pain, that they are capable of learning and intelligent behaviour, and that they probably do share many of the same emotional behaviours as human beings. Darwin adopted an anthropomorphic style with the intention of showing that emotions and expressions were not unique to human beings, while his comparisons between human cultures underlined the universality of emotional expressions.

In making a case for the evolution of language, Darwin suggested that natural selection may act upon entities other than organisms, anticipating Richard Dawkins's (1976) idea of the meme. Darwin wrote that:

> A struggle for life is constantly going on amongst the words and grammatical forms in each language. The better, the shorter, the easier forms are constantly gaining the upper hand, and they owe their success to their own inherent virtue. (1871, p. 60)

At the time that Darwin's works on human behaviour were published, the field of psychology was dominated by physiologists who were investigating the mechanisms of the brain, and by philosophers theorizing about the workings of the mind. To 18th-century British philosophers such as John Locke, David Hume, and John Stuart Mill the human mind at birth is like an empty box, which is free of in-built knowledge and is gradually filled as we experience the world. Eventually, our ideas and observations become integrated so that we can make sense of that around us, an idea that became known as *associationism*. With hindsight, we can see how this idea must be wrong. We cannot construct a mental picture of the world unless we have ready-built structures that make knowledge acquisition possible. The great German philosopher Immanuel Kant made the point that there must be certain preconditions to the human mind that contribute to our conception of the world in his famous *Critique of Pure Reason* (1781), and Kant's insights have been confirmed by a vast array of recent findings from neuroscience, psychology, and artificial intelligence. We now understand that the mental apparatus that allows us to perceive, interpret, and model the world around us is partly a product of our genes. The publication of Darwin's three great works was partly instrumental in bringing about a decline in associationist views within psychology (Boakes, 1984).

In the second part of *The Descent of Man*, Darwin introduced the concept of sexual selection in order to provide an additional explanation for physical and mental differences between the sexes. Following the principles of natural selection, this idea stated that characteristics may have evolved that increase an individual's chances of gaining matings either through enhancing competitive abilities amongst members of the same sex (usually presumed to be more important in males than in females) or through enhancing the likelihood of being chosen as a mate (usually viewed as females choosing particular males). Such factors were suggested to generate selection for particular characters in one or other sex, such as the large antlers of male deer or the peacock's extravagant tail.

Darwin's views on mental differences between the sexes in human beings are now somewhat dated. He wrote that:

> Man is more courageous, pugnacious, and energetic than woman, and has the more inventive genius (p. 316). Male monkeys, like men, are bolder and fiercer than the females (p. 320). These characters will have been preserved or even

augmented ... by the strongest and boldest men having succeeded best in the general struggle for life, as well as in securing wives, and thus having left a large number of offspring. (1871, p. 325)

However, on the down side,

Man delights in competition, and this leads to ambition which passes too easily into selfishness. (1871, p. 326)

Darwin suggested that:

Woman seems to differ from man in mental disposition, chiefly in her greater tenderness and less selfishness. It is generally admitted that with woman the powers of intuition, of rapid perception, and perhaps of imitation, are more strongly marked than in man. (1871, p. 326)

Darwin may be forgiven to some extent if he is compared with the prevailing views of Victorian Britain. His observation that, with education, 'woman should reach the same [intellectual] standard as man' (1871, p. 329) suggests that his views were more liberal than those of many others at the time.

Similarly, although his ideas on racial differences amongst human populations published in *Descent* seem prejudiced by today's standards, he was again willing to consider that opportunity plays a major role in such differences. On his voyage on the *Beagle* from 1831 to 1836, Darwin travelled with three Fuegians from South America (Blackmore and Page, 1989). On a previous voyage by Captain FitzRoy, two had been taken hostage in reprisal for a theft from the ship, while the third had been bought from his parents for a pearl-button, after which he was named. The three had been transported to England, 'educated' into British civilization and Christianity, and were now being returned to their

homeland as intended missionaries. Jeremy Button, in particular, made an impression on Darwin and fellow ship-
mates because of his linguistic abilities, good humour, and
manners. Darwin was stunned when he arrived in Tierra del
Fuego, as the other natives appeared to him 'wretched' and
'wild' in comparison. This brought it home to Darwin that
many differences between peoples were brought about by
climate and culture and that, given the opportunity, mental
development could be fast (Boakes, 1984). As we shall see,
many of Darwin's contemporaries maintained a different
attitude, assuming that apparent sex and race differences in
mental abilities are inevitable and could never be overcome
by enhanced opportunities.

Subsequent careful scrutiny of the private notebooks of
Darwin has revealed that many of the ideas in *Descent* were
conceived as far back as 1838 (Gruber, 1974). His notebooks touched on a wide range of psychological topics,
including memory, learning, imagination, language, emotion, and psychopathology. In his notebook of 16 August
1838, Darwin proclaimed that 'he who understands
baboon would do more towards metaphysics than Locke'.
By this Darwin meant that the study of animal behaviour
would be more useful than philosophy in helping us to
understand how the human mind works. This rather star-
tling claim sounds extraordinarily similar to some of the
bold statements that were to emerge from the field of
human sociobiology a century and a half later, but in
Darwin's case his published work was far more considered
and judicious than his private writings. Nonetheless,
Darwin's view that psychology and the study of human
behaviour should be based on an understanding of biology
and the concepts of variation, heredity, and adaptation did
have an impact.

Galton and the development of eugenics

Darwin's younger cousin, Francis Galton, was one of an inner circle of intellectuals privy to his thoughts prior to *Origin*'s publication. Darwin's emphasis on heredity and individual differences provided a source of inspiration for Galton's work, which sought to explain why people differ in mental ability. His major work, *Hereditary Genius*, was published in 1869. In this book, Galton traced the genealogies of able families amongst the judges of England, the peerage, military commanders, and men of science, literature, poetry, and music. For instance, Galton noted that the Bachs were all tremendous musicians, while Darwin's family (from which he modestly excluded himself) were great scientists. Using this information, Galton suggested that mental abilities were inherited, as opposed to the prevailing view at the time that the human mind acted 'independently of natural laws'.

Galton was a polymath who made major contributions to mathematics, psychology, and evolutionary theory, and pioneered the use of identical twins in the study of genetic influences on behaviour (Forrest, 1974). One of his lesser known accomplishments is the invention of fingerprinting to help the police. He also set up an anthropometric laboratory to undertake the collection of physical and mental testing of men and women. Galton became obsessed with measurement—for example, as he travelled around Britain, he secretly constructed a beauty map of the cities, concluding that the incidence of pretty girls was highest in London and lowest in Aberdeen. He also enlivened dull scientific meetings by attempting to measure the boredom level of the audience, eventually settling on a measure of fidgets per minute, a study which he published in the journal *Nature*.

However, Galton also exhibited extraordinary prejudices (Boakes, 1984). He believed, for example, that some men belonged to the criminal type and that no amount of environmental improvement would alter this, and that the inability of women to distinguish the merits of various wines confirmed the inferiority of female intellectual ability. He tended to ascribe almost all differences between human beings to heredity, what we would now call genes, and virtually nothing to education or opportunity. His hereditarian bias is manifest in his definition of genius, which was 'an ability that was exceptionally high *and at the same time inborn*' (Galton, 1869, italics added). He acknowledged that education could develop the mind's full potential; however, individuals could never rise above their inherited mental capacity. Hence, Galton was opposed to the education and suffrage of women. Galton also exhibited racial predudices; for example, he regarded Africans as having a lower average mental ability than Europeans. Even within Britain, he proclaimed the men and women of southern Scotland and northern England to be of greater worth than those of the midlands and especially London. In the chapter *The Comparative Worth of Different Races*, Galton suggested that 'Every long-established race has necessarily its peculiar fitness for the conditions under which it has lived, owing to the sure operations of Darwin's law of natural selection' (1869, p. 336). Galton maintained that, with time, the more civilized races would inevitably eliminate native races because of the latter's inability to cope mentally with the tasks of the superior civilized society.

Galton's work on *Hereditary Genius* began at around the time that he realized that his wife, Louise, would not be able to have children because of her ill health. This is perhaps one reason why he subsequently became increasingly concerned for the future intellectual quality of humanity, fearing that the lower classes would outbreed the gentry. In an article written in 1894, he stated that 'It has now become a serious necessity to better the breed of the human race. The average citizen is too base for the every day work of modern civilisation' (Forrest, 1974). In *Hereditary Genius*, he earlier stated that 'It seems to me most essential to the well-being of future generations, that the average standard of ability of the present time should be raised' (Galton, 1869, p. 344). To accomplish this, he suggested the active encouragement of early and judicious marriage by those possessing 'favourable hereditary qualities', and for the weak and criminal to be sent to celibate monasteries. Thus arose Galton's theory of eugenics, defined by him as 'the science which deals with all influences that improve the inborn qualities of a race'. While Darwin reported the work of Galton in *The Descent of Man*, he did not totally condone these views, instead stating that the intentional neglect of the weak and helpless would be a 'certain and great present evil' (Darwin, 1871, p. 169).

For Galton, eugenics became a great passion. Ironically, his book *Inquiries into Human Faculty and its Development* (1883), which set out his eugenic ideas, was mainly criticized at the time of publication for its anti-religious views. Using his information on family histories, Galton stunningly concluded the inefficacy of prayer, by showing that men much prayed for, such as those high up in the church, did not live longer than those at the top of other professions such as law, and that ships bearing missionaries sank just as often as those carrying material goods. However, by the turn of the century Galton was regarded as the world's leading psychologist, and his highly hereditarian views were thriving on both sides of the Atlantic (Boakes, 1984).

The ascent of progressive evolution

Towards the end of the 19th century, Darwin's theory of natural selection was losing favour as an explanation for evolutionary change (Bonner and May, 1981). The theory was partly hindered by the lack of knowledge of genetics. However, the main opposition to natural selection came from physicists such as Lord Kelvin. Their calculations seemed to show that the earth was not old enough to have supported life for the thousands of millions of years demanded by natural selection. These estimates are now known to have been incorrect; however, by 1870 the evidence appeared stacked against natural selection. In comparison, the teachings of another great evolutionary thinker, Jean Baptiste de Lamarck, were in the ascendancy.

Lamarck published his works on evolution in 1809 while a professor at the Natural History Museum in Paris. He suggested that all species were independently created and could be placed on a scale with the most similar species next to each other (Blackmore and Page, 1989). Each species could then move up the 'chain of being', which culminated in human beings. The process by which this was thought to occur was the inheritance in offspring of characteristics acquired by parents during their lifetime, such as the passing on of learned knowledge or well-exercised muscles. Lamarck's view of evolution was linear and progressive, with species having an inherent striving to evolve greater complexity, with the pinnacle of creation being human beings. The theory was initially rejected in France, largely due to the opposition of the powerful biologist Georges Cuvier, while in Britain it was regarded as dangerously atheistic and too closely linked with French revolutionary ideas (Boakes, 1984). Indeed, Darwin's emphasis on gradualism was probably partly an attempt to disassociate evolution from revolution. Lamarck died in poverty and scientific disrepute, and at his funeral his daughter is said to have cried out 'My father, time will avenge your memory!' (Boakes, 1984). She was partially right. When the 19th-century physicists stated that there was insufficient time for natural selection to do its work, Lamarck's inheritance of acquired characters seemed to fit, providing a fast evolutionary explanation. While the theory of inheritance of acquired characteristics was eventually proven to be incorrect, the erroneous Lamarckian view equating evolution with progress unfortunately still survives even today.

One advocate of Lamarckian ideas, Herbert Spencer, was particularly influential in the late 19th and early 20th centuries (Oldroyd, 1983). Spencer was born in England in 1820 and, although he initially trained as a civil engineer, his major interests were psychology and philosophy. Spencer cultivated and widely published the idea that all things change inevitably from a simple to a more complex state, including species and human societies. Spencer's concept of mental evolution was that of a single continuum from the reflexes of simple animals to their pinnacle in the intelligence of the civilized man. In his influential *Principles of Psychology* (1855 and 1870), Spencer described how human societies gradually became more developed, with the 'large brained European' mentally far in advance of 'primitives'. A similar view, that human society progresses through various levels punctuated occasionally by revolutions that take a society to a higher level, was being propounded by Karl Marx.

In the United States of America in the late 19th century, Spencer's views of evolution and society rivalled Darwin's for popularity, and were endorsed by religious and business leaders (Oldroyd, 1983). When Spencer travelled to America in 1882, he was warmly greeted and his books were bought by the thousands as his views justified the business ideas of the newly wealthy country. Spencer's slogan 'survival of the fittest' was eagerly accepted by business, where it was quite clear that fitness was to be measured in wealth. This endorsement of evolutionary ideas by society and business began the movement known as 'Social Darwinism'. However, 'Social Spencerism' would be a more appropriate term, since it derives far more from Spencer than Darwin. The title of Spencer's 1894 book, *The Ascent of Man*, indicates the level to which evolutionary thinking had become an all-embracing notion of progress and design just 20 years on from the publication of Darwin's theses on human evolution.

As Social Darwinists erroneously believed that evolution was progressive, they drew the conclusion that it should be encouraged, and used it to justify doctrines such as social conservatism, militarism, eugenics, laissez-faire economics, and unfettered capitalism (Oldroyd, 1983). The leading Social Darwinist among American academic circles was William Sumner, Professor of Political Economy at Yale. Sumner asserted that:

> Millionaires are a product of natural selection ...They get high wages and live in luxury, but the bargain is a good one for society. (Oldroyd, 1983)

In contrast, socialist schemes were regarded as a menace to society as they 'promote the survival of the unfittest'. Business leaders, such as Andrew Carnegie and J. D. Rockefeller, also exploited evolution to their own ends. For example, Carnegie argued that 'the concentration of business in the hands of the few...was essential to the future

progress of the race' (Oldroyd, 1983). This is a gross distortion of Darwinian thought and Darwin wholly rejected
such interpretation of his ideas.

Social Darwinism thrived partly because during the last
two decades of the century the idea that nature counted
much more than nurture in the expression of human
behaviour overwhelmed Europe and North America. The
huge contrast between the power and wealth of these
nations compared to the rest of the world came more and
more to be seen as reflecting in-built differences in the psychology and abilities of different 'races'. For example, Ernst
Haeckel, an eminent and powerful German professor of
zoology, championed the view that the evolution of a
species, like an individual's development, progressed
through increasingly higher stages. He cannonized this idea
with his 'biogenetic law', which suggested that ontogeny
(development from conception to death) is a re-enactment
of phylogeny (the evolutionary history of the species).

Haeckel had been converted to evolutionary thinking on
reading *The Origin of Species*, and was an energetic recruit,
writing a series of papers and books on evolution that
established him as one of the world's leading evolutionists
(Boakes, 1984). However, like Spencer, Haeckel's view of
evolution tended more towards Lamarck's. For Haeckel,
evolutionary theory also had very definite political implications, and provided the framework for his commitment to
the reform of political institutions and to the unification
of Germany. He was anti-semitic, and used his immense
authority in German-speaking countries to promote
views on inherent racial differences right up to the First
World War. Historians have noted a strong biological tradition passing directly from Haeckel to the appalling
doctrines of Nazi theorists (Oldroyd, 1983). Years later,

pseudoevolutionary political diatribe was to reach its dia-
bolical zenith with the publication of *Mein Kampf*, in which
Adolf Hitler drew on facile analogies from animals an erroneous conception of 'blending inheritance', and Spencer's
'survival of the fittest' doctrine to give a quasi-scientific
argument for the need for racial purity. In biological terms
Hitler's arguments were nonsensical, yet no body of work
illustrates more dreadfully how dangerous is the distorted
view of evolution as progress.

Even George Romanes, chosen by Darwin as his suc-
cessor, was to regard evolution in progressive terms
(Boakes, 1984). A strong friendship had developed between
the two men after Romanes began writing to Darwin in
1874. Romanes addressed the question of how human
mental abilities had evolved by comparing human and animal behaviour. The animal mind was an extraordinarily
popular topic in the 1870s, and countless letters flowed into
scientific and popular journals reporting striking observations of animals' mental capabilities. Romanes took to
collecting and examining these anecdotes, and published a
report on them in his 1882 book *Animal Intelligence*. The
treatise collates countless examples of animal champions
and boffins arranged in order of mental ability, from the
earwig that had been trained to climb up the curtain every
day to eat breakfast, to the dog that understood the
mechanical principle of the screw. However, Romanes
appears to have been more influenced by Spencer and
Haeckel than by Darwin, and cited both frequently. Using
Haeckel's biogenic law, Romanes placed the mental abilities
of animals on an ascending scale, culminating in humans.
He then proposed that, during development, a human
being plays out this evolutionary ladder. Each age was
ascribed a comparable animal intelligence (Table 2.1); for

Table 2.1 Romanes' (1882) depiction of the relative intelligence
of humans at various stages of mental development and the
corresponding level of other species

Human development	Equivalent to	Psychological ability
Sperm and egg	Protoplasmic organisms	Movement
Embryo	Coelenterata	Nervous system
Birth		Pleasure and pain
1 week	Echinodermata	Memory
3 weeks	Larvae of insects	Basic instincts
10 weeks	Insects and spiders	Complex instincts
12 weeks	Fish	Associative learning
4 months	Reptiles	Recognition of individuals
5 months	Hymenoptera	Communication of ideas
8 months	Birds	Simple language
10 months	Mammals	Understanding of mechanisms
12 months	Monkeys and elephants	Use of tools
15 months	Apes and dogs	Morality

example, at 3 weeks of age, a baby was roughly equivalent to
an insect in mental ability; at 4 months of age, it was equal
to a reptile; by a year, a child was as clever as an elephant;
and by 15 months, it was usually brighter than apes and
dogs.

The idea that human societies progressed through
various levels was also prevalent within the emerging
field of anthropology (Oldroyd, 1983). Even the anthropologists with whom Darwin was most closely associated,
John Lubbock and Edward Tylor, had no doubt that
higher cultures were associated with more advanced races
whose members had larger and more effective brains.
Lubbock and Tylor argued that all civilized nations are the

descendants of barbarians, first, because some traces still existed in customs and language, and in archaeological remains such as flint tools, and secondly, because 'savages' were sometimes independently able to raise themselves a few steps in the scale of civilization. Tylor set out this theory in his two major publications, *Researches into the Early History of Mankind and the Development of Civilisation* (1865) and *Primitive Culture* (1871). He reasoned that if one studied the stone-age cultures in other parts of the world one could gain historical insights into the past stone-age culture of Europe. In 1877, in his book *Ancient Society, or Researches in the Lines of Human Progress from Savagery through Barbarism to Civilization*, Lewis Henry Morgan took this viewpoint to its logical extreme by documenting the stages of cultural evolution through which societies were assumed to progress (Table 2.2).

These anthropologists argued that all races of human beings shared a common ancestor, but that some races were higher on the scale of progression than others. This view was in contrast to the ideas of another set of anthropologists, who argued that slavery was natural because different races were actually different species (Oldroyd, 1983).

Table 2.2 Lewis Henry Morgan's (1877) stages of cultural evolution

Lower savagery	Fruit and nut subsistence
Middle savagery	Fish subsistence and fire used
Upper savagery	Bow and arrow used as weapon
Lower barbarism	Pottery used
Middle barbarism	Animals domesticated, maize cultivated with irrigation, adobe, and stone architecture
Upper barbarism	Iron tools used
Civilization	Phonetic alphabet and writing employed

This latter group spent their time in the physical description and classification of different races around the world. Such racism relied on the idea that human beings within a population could all be described as a particular type. Yet Darwin's view of evolution crucially highlighted the importance of variation within populations and rejected such typological thinking. The evolutionary evidence clearly supported the single-species view. Thomas Huxley argued that the ability of all humans to interbreed implied we must be one species, and Darwin's work on the similarities between races in mental abilities and expression of the emotions clearly backed this view.

The widespread idea that British and North American society was superior to that of other cultures provided even greater impetus to the Social Darwinist movement (Boakes, 1984). Victorian social institutions were presumed to be natural, good, and healthy, whereas 'primitive' societies were abnormal and degenerate. During the 1890s, a number of biologists, including Thomas Huxley in Britain and James Mark Baldwin in the United States, reacted angrily to what they saw as the damaging use of evolutionary theory to justify obnoxious social and ethical values. Unfortunately their protestations fell on deaf ears.

The nature–nurture debate

The leaders of Victorian society kept a close and informed interest in current scientific developments, including zoology, and consequently publicity was drawn to a set of extraordinary experiments carried out on birds by a young British scientist called Douglas Spalding (Boakes, 1984). These studies inclined many readers to consider that the human mind might depend upon instinct.

Originally earning a living mending slate roofs, Spalding educated himself by attending public lectures on philosophy at Aberdeen and London. He became frustrated that the leading psychologists and philosophers were prepared to discuss whether the mind was, or was not, influenced by instincts without ever testing these assertions. Spalding began to carry out his own set of experiments on young chicks to investigate whether any inherent abilities were present at hatching. To test whether a young chick was able to move about its world without bumping into objects, to peck accurately, and to locate sounds, without any prior sensory experience, he removed sections of shell from eggs just before the birds emerged, put wax in their ears and covered their eyes with a patch to remove any auditory or visual cues, and then tested them after hatching when the wax and hoods were removed. He concluded that these birds were just as capable as other birds of pecking accurately, making coordinated movements, avoiding objects, and responding appropriately to threats like a hawk, and concluded these abilities must be 'instinctive'. He also discovered that chicks would imprint on, or latch onto, the first object that they see after they hatch, which is usually their mother, and would 'instinctively' follow her around.

Spalding was later employed by Lord Amberley, the son of the Prime Minister, as a tutor to his eldest son and was encouraged to continue his pioneering research in their house, with Lady Amberley as his assistant. Unfortunately Spalding's research ended suddenly in scandal. After the deaths of Lord and Lady Amberley, the guardianship of the sons was left to Spalding, to the consternation of their powerful grandfather. The guardianship was fiercely contested and Spalding was forced to emigrate to France, where he died a year later at the age of 37. The philosopher

Bertrand Russell, another of Lord Amberley's sons, later revealed that Lady Amberley had taken Spalding to bed out of a motherly concern for his celibacy (Boakes, 1984).

In Britain, emphasis was being placed more on the comparison between human mental abilities and those of other animals. Conwy Lloyd Morgan opposed Romanes' anecdotal approach to the study of the mind and undertook his own comparative study of instinct. One of the founding fathers of both comparative psychology and ethology, Lloyd Morgan wrote 14 substantial books, including *Habit and Instinct* (1896), *Animal Behaviour* (1900), and *The Animal Mind* (1930). In particular, he propounded the notion of using accurate definitions and observational data, the replication of experiments, and the avoidance of implying complex mental attributes in animals where such abilities are unproven.

Distorted views of evolution continued to influence scientific thinking into the early 20th century, particularly in psychology (Richards, 1987). The American George Stanley Hall, who was one of the founders of psychology as a subject at university and advocated its practical benefits to teaching and raising children, utilized Lamarckian inheritance and Haeckel's biogenetic law as major principles in his work. Sigmund Freud's theories of psychopathology were also greatly influenced by Darwin and Haeckel (Sulloway, 1979; Richards, 1987). Freud took Darwin's ideas of sexual selection and the 'instinct' to mate, and used them to develop his concept of the libido, a core of instinctive urges, chiefly sexual, that were the unstated driving force behind human behaviour. Freud's view that one could gauge the inner workings of the human mind indirectly through what was happening on the surface, and that illnesses might be ascribed to forgotten experiences, was influenced by Darwin's work on the expression of the emotions. Moreover, Freud's psychosexual theory draws directly from Haeckel's discredited biogenetic law (Sulloway, 1979). If animals at the equivalent developmental stage are sexual creatures, Freud reasoned that infant humans must be too, going through an oral stage when they gain sexual pleasure from the mouth, later to be followed by anal and phallic stages.

reds: erotic

One influential psychologist who challenged the Spencerian view of psychology was the American William James (Plotkin, 1997). Initially an admirer of Spencer, James became dissatisfied with the passive and deterministic view of human behaviour that dominated psychology. Instead, James reverted to a more Darwinian perspective, proposing that the mind generated ideas (variation) rather than being shaped passively by the external world, and that those ideas that provided the best way of dealing with the world would be retained (selected). He believed in the importance of adaptation in explaining important features of the mind such as consciousness and instinct. His textbook *Principles of Psychology*, first published in 1890, ran to several editions. William McDougall, another eminent Harvard professor, argued that animals should be studied in order to understand the core human nature, characterized by the emotions and instinct.

In contrast, James Mark Baldwin, founder of the *Psychological Review*, the premier psychological journal, and architect of the Baldwin Effect, adopted an evolutionary approach to psychology, but rejected simple hereditarian views of human behaviour (Boakes, 1984). Baldwin endeavoured to develop psychological principles consistent with evolutionary theory, but which none the less accounted for the influence of cultural inheritance. His major interest was child development, a topic which Baldwin believed had been distorted by the genetic approach of George Stanley Hall. Through careful observation, he charted the gradual appearance of the different mental powers in human infants, determined the sequence in which they emerge, and emphasized the possible importance of imitation in mental development. Tragically, this clever man, one of America's leading psychologists, was forced to resign from Johns Hopkins University in 1909 after being arrested in a brothel *Baldwin scandal* (Boakes, 1984). While Baldwin proclaimed his innocence, his abrasive style had won him few friends in academic circles, and his contributions to psychology were written out of the history books. Nonetheless, after moving to Paris in disgrace, Baldwin continued to have an influence on psychology, particularly through the Swiss psychologist Jean Piaget.

Reaction to any instinct-based theories of human behaviour and to eugenics gathered momentum in the early 20th century. Part of this dissatisfaction was that the concept of instinct was increasingly criticized as being vague and unscientific. One review reported that in the previous twenty years nearly six thousand types of instinct had been proposed, including the instinct of girls to pat and arrange their hair, and the desire to liberate the Christian subjects of the Sultan (Boakes, 1984).

At the beginning of the First World War, the US army had allowed the psychologist Robert Yerkes to carry out intelligence testing on the forces with a view to improving the intake and efficiency of recruits, and nearly two million men had been tested. After the war, when the tests were analysed, the theoretical assumptions made by Yerkes and colleagues were strongly hereditarian (Boakes, 1984). The results suggested that intelligence varied with race and,

Yerkes .

among immigrants, those that had most recently moved to the States performed worse than those of families with longer residence in the country. These data were taken as proof that, as widely feared, the mental calibre of immigrants had been steadily declining. A more likely explanation is that, as immigrants would become increasingly familiar with American culture over time, those immigrants with longest residency would score better on the tests, but this was ignored. While, in the 1920s and 1930s there was mounting criticism of the use of intelligence testing, President Coolidge was among those who accepted Yerkes's conclusions, and as a result he imposed an Immigration Act in 1924 that restricted immigration to favoured races and nationalities. Fifty years later, the restrictive immigration laws were to be cited by the critics of human sociobiology as a prime example of the dangers of evolutionary methods applied to human behaviour.

Within psychology, there was a shift in emphasis towards studying only those behaviour patterns that could be observed and measured. The predictability and control of behaviour such as reflex actions and stimulus-response learning became the focus of attention, with the study of learning being the central theme. This school of thought, known as behaviourism, began with the publication of works by John Watson in 1913. Watson rejected the notion that inheritance played any meaningful part in explaining human behaviour. He stated that we need only consider what is learned to understand human behaviour, and that learning is the proper focus for psychology. In a well-known quotation, Watson (1924) boldly claimed:

> give me a dozen healthy infants, well-formed, and my own
> specified world to bring them up in and I'll guarantee to
> take anyone at random and train him to become any type

of specialist I might select—doctor, lawyer, artist,
merchant-chief and, yes, even beggar-man and thief,
regardless of his talents, penchants, tendencies, abilities,
vocations, and race of his ancestors. (Boakes, 1984)

Behaviourist psychology in the United States conformed
better with the political ideology that stressed equality of
opportunity. A parallel movement had developed in Russia
based on the research of the physiologist Ivan Petrovich
Pavlov (Boakes, 1984). Lenin is said to have paid a secret
visit to Pavlov's laboratory in 1919 to find out if Pavlov
could help the Bolsheviks control human behaviour
(Bateson and Martin, 1999). Pavlov told him that 'natural
instincts' could be abolished by a form of learning now
known as 'Pavlovian conditioning', a view so congenial to
Lenin that it became the party line, and Pavlov's research
was widely promoted. By the 1930s, the idea of instincts had
largely disappeared from experimental psychology.
Evolution was the baby that went out with the bath water.

Shortly after the rise of behaviourism in psychology, a
similar reaction against instinct and hereditarian views
occurred within anthropology (Boakes, 1984). The leader
of the new movement was Franz Boas. In 1883, as a 25-year-
old student in Berlin, Boas went to live among the people of
Baffinland and became aware of a relativity and arbi-
trariness to human customs. This was strengthened by
expeditions to study the Indians of British Columbia. Boas
did not deny the parallels across cultures, but disputed
whether they implied a universal sequence of development.
Boas therefore urged careful study of individual cultural
communities and avoidance of the overarching generaliza-
tions of the evolutionist school. Together with his students
Margaret Mead and Ruth Benedict, Boas pioneered a new
anthropology dominated by the ascendancy of nurture over

nature, arguably as extreme as the evolutionist movement. Culture was thought to determine social life completely— even the most basic elements of how we mate and bring up our children was thought to be constructed by cultures and differ from one place to another. Perhaps Boas and co-workers were endeavouring to counter the rise of racist views among Social Darwinists. The relatively swift transition from hereditarianism to environmentalism in the 1930s was in part due to the efforts of Franz Boas, Margaret Mead, and Ruth Benedict.

Ironically, psychology, anthropology, and the other human sciences rejected evolution at precisely the time that evolutionary theory was really coming together. The modern synthetic theory of evolution was forged in the 1930s, with the integration of Mendel's genetics and Darwinian thought, the rejection of Lamarckian inheritance, and with natural selection re-established as the major evolutionary process. The classic works of Theodore Dobzhansky (*Genetics and the Origin of Species*, 1937), Ernst Mayr (*Systematics and the Origin of Species*, 1942), Julian Huxley (*Evolution: the Modern Synthesis*, 1942), and George Simpson (*Tempo and Mode in Evolution*, 1944) showed how the new Synthetic theory could be employed to make sense of evolutionary lineages and of the characters of contemporary populations of organisms. Evolutionary theory gained a solid theoretical foundation through the works of J. B. S. Haldane, R. A. Fisher, and Sewell Wright in the 1920s to 1950s, in which the methods of population genetics and the mathematical theory of evolution were worked out, and key concepts such as fitness defined. Evolutionary biology could now be regarded as a mature science.

One scientific development that resulted directly from the emergence of the modern Synthetic theory of evolution

was the need to catalogue genetic variation in natural populations. This research soon revealed that genetic differences between human populations are small compared
with the great amount of variation within them. This data
supported the vigorous arguments that many evolutionary
biologists, including Dobzhansky (1962), were making
against racism.

Ethology and the resurrection of instinct

As the majority of psychologists and anthropologists dis-
regarded evolutionary arguments, an increasingly persua-
sive body of knowledge and valuable new set of methodologies for the study of behaviour were being developed. This
science became known as ethology, from the Greek 'ethos'
meaning character (Thorpe, 1979). Using a knowledge of
the natural history of animals, the ethologists set out to
examine the robust behaviour patterns that are seen within
one species and not another. The idea of instinctive behaviour was once again re-emerging. Here, we spend some
time reviewing ethology, as this work provides much of the
background for the fields that we describe in the rest of the
book. In the next section, we describe some of the work on
human behaviour that was carried out under the name of
ethology, some of which was actually damaging to its scientific reputation.

  In the late 19th and early 20th century, two scientists,
Oskar Heinroth in Germany and Charles Otis Whitman in
America, were independently documenting patterns of
movements, such as courtship behaviour in birds
(Burkhardt, 1983). Heinroth observed that the precise
movements of ducks engaged in courtship was highly characteristic of a species and that the similarities and differ-

ences between species could be used in exactly the same way as physical characteristics to trace common ancestry and reconstruct the evolutionary past. Whitman's studies of pigeons led to similar findings. A few decades later, a young Austrian anatomy student called Konrad Lorenz, heavily influenced by this work, came to the conclusion that the methods employed in comparative morphology could be applied to the behaviour of animals. Lorenz was determined that 'the phylogenetic view' (his term for an evolutionary perspective) should triumph in the study of animal behaviour. From early childhood, Lorenz had an 'inordinate love of animals' and, knowing nothing of Spalding's work, had independently discovered imprinting through his experiences hand-raising a flock of geese in his home village of Altenberg (Wasson, 1987). The picture of Konrad Lorenz being followed around the Austrian countryside by a line of young goslings has become one of the most enduring images of ethology. In 1936, Lorenz met Nikolaas Tinbergen, a zoologist at the University of Leiden, in Holland, who had developed a research programme characterized by the observational and experimental study of animals in their natural environments. They were amazed at the similarities of their views, and struck up an immediate friendship.

Lorenz and Tinbergen were the true founders of ethology, and pioneered a novel approach to the study of behaviour (Hinde, 1982). By the early 1950s, ethology had emerged as a new discipline, with Lorenz as its father figure and *The Study of Instinct* (1951) by Tinbergen its classic text. The elegant studies of another great Austrian ethologist, Karl von Frisch, on communication in the honey bee, are arguably ethology's most famous insights. Ethology also flourished in England, with Bill Thorpe and Robert Hinde at Cambridge pioneering the study of birdsong and behavioural development in birds and primates, and in Oxford following Tinbergen's move to England in 1950. Ethology also made an impact in the United States by the middle of the century, particularly through the work of William Morton Wheeler and Karl Spencer Lashley (Thorpe, 1979).

The ethological method typically began with an extensive period of observation of the animal in its native environment, followed by a careful description of the relevant behaviour patterns, known as an 'ethogram'. A variety of stereotypical behaviour patterns or *fixed motor patterns* for that species were identified. Lorenz suggested that the tendency to produce an instinctive behaviour built up over time and, when activated by the appropriate stimulus, found expression in a fixed motor pattern. Some ethologists tried to explain instincts in physiological terms, in a manner that was subject to experimental investigation. To the ethologists, instinct was an inherited and adapted system of coordination within the nervous system. In addition, Tinbergen in particular focused on the survival value of particular behaviour patterns and was exemplary at designing simple experiments that would test causative and functional hypotheses in natural conditions.

Ethology was constantly engaged in a running battle with the American school of comparative psychology, which arose from behaviourism. The two groups shared an interest in animal and human behaviour, but they approached it from very different viewpoints. The ethologists worked largely in Europe and, being biologists and naturalists, they largely studied animals in their natural environments. In contrast, and despite their name, the comparative psychologists were not concerned with comparisons between species, but tended to focus on just one or two species, such as rats or pigeons. This is because they believed there were general rules of behaviour that would hold regardless of the species being studied and the experimental context. The ethologists maintained that the psychologist's so-called general rules were artefacts of the impoverished experimental conditions.

An important critique of ethology was written by the American psychologist Daniel Lehrman in 1953. Lehrman dismissed ethologists' accounts of innate behaviour, first, because organisms never develop in complete isolation from their environment and therefore one could never know that a behaviour pattern was uninfluenced by external events, and secondly, because 'innate' was defined in terms of excluding what is learned, it would never be a usable concept. Earlier, T. C. Schneirla had suggested that the relative importance of 'innate' and 'acquired' effects on behaviour patterns could not be separated and that an individual's development is a complex interaction of genetic information, the developing organism, and its environment (Hinde, 1982).

Perhaps because the ethologists were so preoccupied with their battle with the comparative psychologists, they constantly stressed the characteristic fixed behaviour patterns of a species and neglected how individuals vary within a species. That variation was central to Darwin's perspective. Lorenz's early training in comparative anatomy may have accounted for his typological thinking, and his influence may help to explain why many ethologists repeatedly made the mistake of thinking that natural selection was a process that operated for the good of the species. If all individuals are thought to behave in the same manner, then it is easy to envisage that their interests are aligned. Eventually, the ethologists conceded that 'instinct' was not an adequate explanation for human behaviour, not least because it discouraged interest in behavioural development (Hinde, 1982). This realization led Tinbergen (1963) to add 'how does a behaviour develop?' to the three questions of biology outlined by Julian Huxley; namely, what are proximate physiological causes, the function (or survival value), and the phylogenetic (or evolutionary history) of a behaviour? Ethology had identified four important classes of question that can be asked about behaviour, which Robert Hinde (1982) was later to label 'core ethology'.

One of Lorenz's major contributions to the understanding of animal behaviour was his view that learning itself is an evolved ability, and that both instinct and learning are of importance and not mutually exclusive. In his 1965 book, *Evolution and Modification of Behavior*, Lorenz provided a partial solution to the nature–nurture debate that has generally been overlooked, introducing the concept of the innate 'school marm' that instructs learning. One of the most important contributions of ethology to the social sciences is the idea that the development of an individual is channelled but not predetermined, with evolved predispositions influencing when, what, and how an animal learns.

Human ethology

In 1972, Lorenz, Tinbergen, and von Frisch were awarded the Nobel Prize for Physiology or Medicine 'for their discoveries concerning the organization and elicitation of individual and social behaviour patterns'. That the first Nobel Prize to be awarded for the study of behaviour and the causes of behaviour went to ethologists caused great discussion and dispute amongst the psychologists. However, the award reflected the optimism current at the time that work in ethology would generate new understanding in medicine and psychiatry, and shed light on human behaviour.

From the outset, Lorenz believed that ethology would furnish important insights into human behaviour. In virtually all of his popular books the final chapter reveals what the preceding pages have to say about humans. However, his views on human behaviour were tarnished by politics. In the early 1940s, Lorenz wrote thinly veiled scientific papers that are commonly interpreted as supporting the Nazis, their ideal of racial purity, and the selecting out of so-called degenerate elements in society. Many years later, Lorenz confessed that he had found some of the Nazi theories attractive but had been politically naive, and had no conception that they would result in genocide (Evans, 1975). It was only late into the war that he realized the evil of Nazism. Nonetheless, the reader of these articles would not find it difficult to understand why Lorenz's critics would charge him with abusing biological arguments to justify racism. In contrast, Tinbergen's experiences in occupied Holland and in a hostage camp, left him unable even to bear the sound of spoken German.[1] The Second World War delayed the development of ethology, cutting off relationships between colleagues and friends, although Lorenz and Tinbergen were to renew their friendship many years later.

Lorenz's 1963 book *On Aggression* caused a major furore, and greatly upset many intellectuals and social scientists (Salzen, 1996). Lorenz argued that fighting and war are the natural expression of human instinctive aggression, which, according to his theory of instincts, inevitably wells up in us unless otherwise expressed, and would be discharged spon-

---

[1] Robert Hinde, personal communication.

taneously and without reason. Despite a final avowal of optimism, Lorenz paints a bleak picture:

> An unprejudiced observer from another planet, looking on man as he is today, in his hand the atom bomb, the product of his intelligence, in his heart the aggressive drive inherited from his anthropoid ancestors, which this same intelligence cannot control, would not prophesy long life for the species. (1966, p. 40)

In typically blunt terms, Lorenz suggests that attempts to eliminate aggression by appropriate training, or by shielding human beings from all circumstances that might elicit it, 'have no hope of success whatever' (1966, p. 239). Lorenz argued that the only chance for humanity is to face up to the grim reality, charging us 'know thyself', and suggesting one or two rather uncompelling solutions, such as to engage in more sport to release aggressive urges.

Lorenz's book provoked considerable hostility (Salzen, 1996), and was disowned by many English-speaking ethologists.[2] Critics objected to his extrapolation from animals to humans, many argued that aggressive behaviour was learned, and others drew the disturbing conclusion that if aggression was the expression of an inescapable urge then war is unavoidable (Salzen, 1996). The opposition and debate continued for more than twenty years. In 1983, a group of expert scientists met at a meeting on aggression in Spain, and drew up what has become known as 'The Seville Statement on Violence' (presented in Table 2.3). Endorsed by major professional bodies and published in prestigious journals, the statement was eventually adopted and disseminated by UNESCO, with the express purpose 'to dispel

[2] Ibid.

Table 2.3 The 1986 Seville Statement on Violence

1. It is scientifically incorrect to say that we have inherited a tendency to make war from our animal ancestors
2. It is scientifically incorrect to say that war or any other violent behaviour is genetically programmed into our human nature
3. It is scientifically incorrect to say that in the course of evolution there has been a selection for aggressive behaviour more than for other kinds of behaviour
4. It is scientifically incorrect to say that humans have a violent brain
5. It is scientifically incorrect to say that war is caused by instinct or any single motivation

the widespread belief that human beings are inevitably disposed to war'. Ironically, Lorenz makes none of the 'scientifically incorrect' statements of the Seville tract, although he comes close. As we shall see in subsequent chapters, Lorenz would not be the last person to have the critics create a straw-man version of his evolutionary arguments to destroy.

Lorenz was far from the only ethologist to address human behaviour. An entire sub-discipline of 'human ethology' emerged in due course (e.g. Cranach *et al.*, 1979). Tinbergen, in retirement, spent many years using ethological methods to study early childhood autism and stress-related diseases. The psychoanalyst, John Bowlby, greatly influenced by Robert Hinde, adopted an ethological perspective to help explain why young children become attached to their mothers, and why they experience great anxiety when deprived of this contact. Lorenz's student, Iranaus Eibl-Eibesfeldt extended Darwin's study of emotion by travelling round the world photographing the facial expressions of people from different races expressing particular emotions, including aboriginal people with little previous contact with the outside world. In the States, Ekman carried out similar studies, each concluding that the same facial expression represents the same feeling all round the world. These ethologists stood up against the anthropologists such as Mead, who viewed expressions as being culturally determined. However, human ethology did not live up to this early promise, perhaps because many ethologists themselves recognized the need to take account of the peculiar complexities of human beings (Hinde, 1982; 1987). While ethological concepts and methods were assimilated into many other disciplines, within the study of animal behaviour itself ethology was overtaken by the emergence of the new field of sociobiology. The focus turned away from cause and development of behaviour that had been stressed by ethologists towards questions of function and evolution.

The scientific credibility of applying ethological methods to studying human behaviour was additionally damaged by the popularized version put forward by Desmond Morris, a zoologist and the curator of mammals at London Zoo. In 1967, Morris created an even bigger controversy with the publication of *The Naked Ape* than Lorenz had with *On Aggression*. It was an extraordinarily popular book that was to sell well over 10 million copies and be translated into every major language. The basic premise was that humans can best be understood as typical primates that turned to hunting. 'His whole body, his way of life, was geared to a forest existence, and then suddenly...he was jettisoned into a world where he could survive only if he began to live like a brainy, weapon-toting wolf' (1967, p. 16). Morris argued that 'the fundamental patterns of behaviour laid down in our early days as hunting apes still shine through all our affairs' (1967, p. 26), and went on to provide unsupported evolutionary explanations for our sexual behaviour, parental behaviour, aggression, and virtually every other aspect of our daily lives.

Morris depicted himself as a simple ethologist describing the human animal in honest zoological terms, giving readers straight biological truths about the animal selves they had been loath to contemplate. However, the flowing prose was rife with sex and sensationalism, and he frequently touched on sensitive topics. For instance, Morris (1967) stated that pornography and prostitution are 'comparatively harmless and may actually help' (p. 63), that women are wrong to stop their husbands going out with the boys (p. 128), and warned that if women take on masculine traits they risk making their sons homosexual (p. 66). Many fellow ethologists understandably did not approve of Morris' writings.[3] Lorenz stated that he didn't agree with some aspects of *The Naked Ape* because it treated humans as if culture was a biologically irrelevant phenomenon (Evans, 1975).

In the 1960s and early 1970s, there was a proliferation of popular books that, like Morris' writings, built on ethological arguments to postulate a human nature rooted in an earlier primate or hunter–gatherer existence, and thereafter set out to explain a number of aspects of current social behaviour as reflections of our evolutionary past. Other books in this genre included Robert Ardrey's (1966) *The Territorial Imperative*, Lionel Tiger's (1969) *Men in Groups*, and Tiger and Robin Fox's (1971) *The Imperial Animal*. Commonly, such books excited controversy as the descriptions of purported 'innate' behavioural tendencies were seen as justifications for existing social inequalities (Segerstråle, 2000).

---

[3] Ibid.

A history of sense and nonsense

With hindsight, we can now see that books like *The Naked Ape* are representative of a long line of texts, dating back to those of Darwin's contemporaries, that use evolutionary arguments to tell the reader what is 'right', 'natural', or 'inevitable'. From the beginning, self-appointed evolutionary evangelists have been serving up biological 'home truths', while others, such as Thomas Huxley, have objected to the more excessive claims and suggested that prejudice and ulterior motive lie behind their conjecture. Little wonder, then, that many people are wary of evolutionary arguments.

We can also see that, historically, certain ideas have tended to go together: a Lamarckian view of evolution, with species arranged on a ladder and a linear, progressive concept of change, perhaps inevitably engenders prejudice as some evolved forms must be regarded as more advanced, or 'higher', than others. Many of the inequitable views on human races indirectly resulted from this Lamarckian viewpoint. In contrast, the Darwinian conception of evolution stresses within-species variation and rejects the typological thinking that is inherent in racism. In addition to the role of natural selection, modern Darwinism places considerable emphasis on chance events such as mutation and genetic drift. There is nothing about natural selection that supports a progression of populations towards an end goal or 'higher' state. In fact, the misrepresentation of evolution as progressive was so apparent to Darwin that in his notebooks he reminded himself to 'never say higher or lower' (Gruber, 1974), and evolutionary biologists now recognize that it is impossible to define any non-arbitrary criteria by which progress in evolution can be measured (Futuyma,

1986). As no variant can be regarded as more advanced than others, Darwinian evolution is inconsistent with racism and Social Darwinism. It is largely by distorting Darwinian thinking that evolution has been used to justify prejudice and inequality. Most of the negative features sometimes unfairly attributed to evolution, including prejudice, racism, sexism, genetic determinism, and Social Darwinism, do not come from Darwin but from others who twisted his theory.

Another characteristic of Darwin's work from which we can learn was his care and diligence in accumulating as much evidence as possible on the subject of his investigations. He finally published *The Origin of Species* twenty years after the idea of natural selection had first sparked his imagination, and it was more than a decade before Darwin said anything substantive about human evolution. Darwin's books are overflowing with evidence and illustrative examples, which are painstakingly weighed up in support of his hypotheses and to refute alternative explanations. This may be contrasted with other works that we have mentioned in this chapter which make bold statements based on little supportive evidence. By the end of this book, we will see that the most compelling evolutionary explanations of human behaviour are those backed up by rigorous accumulation of data, ideally from a large number of sources. Darwin was also highly aware of how society would respond to his work and its possible implications, and took care to build a water-tight case before making his views known to the world.

Although Darwin was not always right in every respect, his idea of evolution by natural selection has withstood the test of time. Over the years, evolutionary reasoning has made invaluable contributions to understanding of topics such as the relationship between learned and inherited traits, the causes of individual differences, and the development of behaviour. It has also led to the rejection of both genetic determinism and the *tabula rasa* view that human behaviour is infinitely malleable. The investigations of the biological basis of imprinting by ethologists have been at the forefront of research into learning and memory, and have led to a new comprehension of how behavioural aspects of development can be linked to an understanding of brain mechanisms. Additionally, evolutionary research has contributed to the debate against racism by showing that genetic variation within populations swamps differences between populations. Yet, for many, such achievements are overshadowed by the negative uses of evolutionary reasoning.

However, recent times have furnished fresh evolutionary insights and new methods that, if used correctly, promise to lend a new impetus to the quest to understand human behaviour and society. These ideas will be introduced over the next few chapters. We begin with the sociobiological revolution, where we see that the controversy that surrounded Lorenz's and Morris's writings was nothing compared to the fracas over human sociobiology.

CHAPTER 3

Human sociobiology

When Lorenz, Tinbergen and von Frisch collected Nobel prizes for their contributions to the study of animal behaviour in 1973, the field of ethology was already starting to be overshadowed by the rise of a new discipline within evolutionary biology. The new approach, known as sociobiology, built on the work of the ethologists but laid much more emphasis on the functional significance of behaviour (questioning why animals have been selected to behave in particular ways) at the expense of causal processes (for example, investigating what stimuli elicit specific behaviour patterns).[1] Sociobiology brought with it a suite of novel methods and insights, and initiated a radical overhaul of evolutionary thinking in the context of animal behaviour. While in Britain and the rest of Europe the transition from ethology to sociobiology may have been more gradual, in the United States of America the new field took off suddenly following its synthesis by Edward O. Wilson,

---

[1] Recently there have been signs that sociobiologists are returning to an emphasis on causal mechanisms. In the fourth edition of their *Behavioural Ecology* (1997, p. 5) John Krebs and Nick Davies write 'In 1975, Wilson predicted the demise of ethology, with mechanisms becoming the domain of neurobiology, and function and evolution the domain of sociobiology. This prediction was fulfilled until recent years, when there has been a welcome renewed interest in linking mechanism and function.'

perhaps because ethology was less prominent there (Kuper, 1994). By the spring of 1976, entire courses were being offered on sociobiology at major universities in the United States, and by the end of the decade several new scientific journals concerned with sociobiological issues had been created. All of a sudden, eager researchers had a fresh methodology, a new set of questions, and the spring of optimism in their step. Imagine the excitement. Puzzles that had taxed the minds of great thinkers, including Darwin, just seemed to be coming into focus under the powerful resolution of sociobiology's tools. Why then, when the behaviour of ants, gulls, and monkeys seemed to fall suddenly into place, should these new methods not be applied to our own species?

The pioneers of this new way of thinking were George C. Williams, Robert Trivers, William Hamilton, and John Maynard Smith. However, two books brought the attention of the general public to the ideas behind sociobiology. In 1975, Harvard professor E. O. Wilson's *Sociobiology: the New Synthesis* made an immediate impression, resulting in a storm of controversy soon after publication. Wilson's important contribution was to create and name the field of 'sociobiology' by showing its scattered practitioners that it existed, and to demonstrate its feasibility and importance (Segerstråle, 2000). A year later, Oxford zoologist Richard Dawkins brought out *The Selfish Gene*, arguably the most popular scientific book of the twentieth century. These books were a celebration of the 'gene's-eye view', the notion that if we wish to understand what characters ought to evolve it is a convenient and useful heuristic to look at the problem from the perspective of the gene and ask which traits would be most likely to increase its frequency in the next generation. Both books successfully captured the potential and excitement provided by the novel ideas and methods that collectively had reinvigorated evolutionary biology, and it is impossible to overstate their impact. Biologists all round the world started rewriting their lecture courses around these two monographs and lay-people were able to comprehend complex ideas being discussed in evolutionary biology.

While Dawkins was careful to distance himself from direct applications of sociobiological methods to humans and argued that culture took humans into a new realm, Wilson, a scientist renowned for the courage of his convictions, was certainly not shy of this challenge. In the final chapter of his book, Wilson turned his thoughts to human nature, offering bold and speculative evolutionary hypotheses for controversial topics such as gender roles, aggression, and religion. He stated quite openly that one of the goals of sociobiology was to 'reformulate the social sciences in a way that draws these subjects into the Modern [evolutionary] Synthesis' (1975, p. 4). Wilson's book was to catalyse the appearance of a stream of works utilizing and extending the theme of human sociobiology. For other researchers, emboldened by a revolutionary zeal, human behaviour had the appearance of rich, easy pickings. The result was a land rush of biologists into the territory of the human sciences, where they received an extremely hostile reception. The unprecedented tumult over sociobiology was to prove the biggest scientific controversy of the decade.

There was, of course, far more to the development of sociobiology than the matters that have concerned its bearing on humanity. Nonetheless, in this chapter we will take an anthropocentric look at sociobiology by reviewing its principle ideas and methods, providing examples of the application of these methods to our own species and then discussing the numerous criticisms that were presented against this research programme.

Key concepts

Wilson described sociobiology as 'the systematic study of the biological basis of all social behaviour' (1975, p. 4), but this all-encompassing statement captures little more than the breadth of Wilson's vision. Wilson synthesized a new discipline by drawing together experimental and theoretical studies of animal demography, population biology, communication, grouping behaviour, parenting, and aggression, in species ranging from micro-organisms through invertebrates to birds, mammals, and finally human beings. By 1975, developments in evolutionary theory and ecology had led to their convergence in a more rigorous theoretical evolutionary framework for the study of animal behaviour. What set sociobiology apart from ethology was the use of a set of key conceptual tools, including the gene's-eye view, kin selection, and reciprocal altruism. Optimality models were also particularly central to Wilson's synthesis (these will be described in more detail in the next chapter), while game theory and evolutionary stable strategies received considerable attention through the writings of Dawkins.

Some of the advances arose in response to the idea of 'group selection'. Prior to the advent of sociobiology as a discipline, little attention was being paid to the question of whether selection was acting at the level of the individual organism, the group, or the species. Most ethologists had not dwelt on this issue and many presumed that individual organisms were selected to behave for the good of the species. The innovative arguments set against this group selectionist view were to lead to important advances in the study of animal behaviour. In this section we present an introduction to some of these ideas and methods, and illustrate how each was applied in a striking and controversial way to interpret our own species.

The gene's-eye view

Advocates of group selection had maintained that many aspects of the social behaviour of animals could be explained by the idea that animals made sacrifices for the good of the group. For instance, some ethologists had suggested that animals would forgo mating or even commit suicide in an attempt to limit their population size, thereby avoiding overexploitation of their food supplies which might lead to a population crash. This view of evolution was most forcefully brought together by a Scottish ecologist, V. C. Wynne-Edwards, in his book *Animal Dispersion in Relation to Social Behaviour* (1962). Wynne-Edwards argued that limitation of population growth could be achieved by some individuals altruistically restraining their reproduction, and thereby provided an explanation for why subordinate individuals within populations often do not breed. Under such circumstances, groups of individuals, or species, that limited their reproduction might be more likely to thrive than groups that overexploited their habitats. Many animal vocalizations, displays, and aggregations were thought to be means by which individuals could assess population density so as to influence their decision on whether or not to reproduce. Similarly, in *On Aggression* (1966), Lorenz described highly restrained and ritualized disputes between animals, fought according to some equivalent of the Queensberry rules that govern boxing. He argued that these should be seen as competition between individuals to determine who had earned the right to breed and who should withdraw, forging contracts that would be favourable for the future of the species.

While these explanations of animal behaviour seemed superficially plausible, the phenomena explained by Wynne-Edwards and others in terms of group selection could be more parsimoniously explained as individuals attempting to maximize their own reproductive success. In 1964, John Maynard Smith published a short rebuttal of group selection and, in 1966, David Lack challenged Wynne-Edward's group selectionist interpretations of the empirical evidence, particularly those on bird populations. However, the most powerful platform against group selection was provided by George C. Williams in his classic 1966 book *Adaptation and Natural Selection*. Williams was highly dissatisfied by group selection arguments. He pointed out that group selection was unlikely to occur where individuals would be able to cheat the system for their own benefit, as such cheaters would out-compete other members of the population and increase in numbers at the expense of others in the group. He also pointed out that the movement of individuals between groups would erode group differences and weaken group selection further. Williams convincingly demonstrated that a simpler and more plausible explanation comes to light if one drops down a further level from the individual and thinks about what characteristics a gene would need to have to increase its representation in the next generation. Williams stated that a 'gene is selected on one basis only, its average effectiveness in producing individuals able to maximize the gene's representation in future generations' (1966, p. 251).

The social behaviour described by Wynne-Edwards in group selection terms, for example the lack of breeding by individuals in poor condition or low in the social hierarchy, could instead be explained in terms of natural selection acting within groups. For example, a gene that increased the probability that its carrier would delay breeding if the individual was in such poor condition that it would only be wasting time and resources might have a selective advantage over a gene that encouraged such an individual to attempt to breed under all circumstances. Similarly, disputes over territories may be understood as competition for the resources required for breeding, and losers may not be able to breed or may be better off not attempting to breed rather than altruistically refraining for the population's sake. Later, this gene-centred perspective was still more powerfully expressed in Dawkins's *The Selfish Gene*. The importance of taking the gene's-eye view will become evident as we discuss the ideas of kin selection, parent–offspring conflict, and reciprocal altruism.

Kin selection

The main difficulty facing evolutionary biologists opposed to group selection was to explain altruism. Why should an individual behave in a way that decreases its own chances of surviving and reproducing and increases another individual's reproductive success? How could such apparently self-sacrificial behaviour have evolved? For example, in many colonies of ants, bees, and wasps (the Hymenoptera), the majority of individuals, known as the workers, are not able to reproduce at any point in their lifetimes and instead devote their efforts to raising the offspring of one or more reproductive females, the queens. In *The Origin of Species*, Charles Darwin described the presence of these workers as 'the one special difficulty, which at first appeared to me insuperable, and actually fatal to my whole theory' (1859,

p. 257). This conundrum had been puzzling evolutionary biologists for over a century. It was not until 1964 that a British graduate student called Bill Hamilton finally devised a satisfactory solution that was consistent with modern genetics: the answer was kinship. Close relatives share copies of many of the same genes, and hence individuals may increase the frequency of these common genes in the next generation by helping closely related kin to reproduce.

Hamilton based his work on that of R. A. Fisher, who had retired from the Department of Genetics in Cambridge in 1957, around the time that Hamilton started his undergraduate studies (Segerstråle, 2000). Hamilton's lecturers at Cambridge and London were mainly group selectionists who disapproved of Fisher (Segerstråle, 2000). Another leading British population geneticist, J. B. S. Haldane, had proposed a group selectionist model of altruism, but Hamilton had quickly rejected it. Hamilton's solution was based on another of Haldane's ideas. In a popular journal published in 1955, Haldane had joked that he would lay down his life for two brothers or eight cousins. In other words, his willingness to forfeit his life would depend upon the benefit gained by his kin and their relatedness to him. Hamilton devised a method for predicting when altruistic behaviour is likely to be selected, depending upon the degree of genetic relatedness between the two individuals involved. Hamilton's theory, which was to become known as the theory of kin selection, was to revolutionize our understanding of animal social behaviour.

The basic idea of kin selection is straightforward. Consider the example of an individual that behaves altruistically to a relative at some cost (denoted as c) to its own life prospects, but that the act benefits (denoted as b) a relative's chances of survival and reproduction. If the propensity to act altruistically is increased by genes that are also present in that relative, then, although the altruist's chances of passing on the genes directly are decreased, the likelihood that the relative will do so is enhanced. Selection of this behaviour will occur if the fitness cost to the altruist is less than the benefit to the relative multiplied by the probability that the relative possesses the same gene (r), or $c<br$. We will see that the importance of weighing up the costs to the donor and benefits to the receiver is central to many of the key concepts in sociobiology.

Robert Trivers · was later to describe kin selection as 'the most important advance in evolutionary theory since Darwin' (1985, p. 47), while Wilson regarded it as 'the most important idea of all' (1994, p. 315). Astoundingly, Hamilton initially struggled to get his theory across and had considerable trouble getting approval for his Ph.D. thesis on the genetics of altruism. Eventually, he managed to publish his work in 1964 as two papers entitled 'The genetical evolution of social behaviour'. The first paper presented the theory, while the second applied the models to the presence of non-reproductive workers in colonies of social insects, the very problem that had confounded Darwin. Subsequently, the American George Price, a brilliant self-taught maverick, showed that kin selection could be regarded as a special case of group selection (1970), a formulation that Hamilton immediately embraced (1970).

Kin selection is of particular relevance to the study of social Hymenoptera because of their unusual form of sex determination, known as haplodiploidy. Female offspring develop from fertilized eggs and therefore have two sets of chromosomes (they are diploid), while male offspring develop from unfertilized eggs and therefore contain only a single set of chromosomes (they are haploid). Daughters will receive identical sets of genes from their father, as he only has one set to give. The other half of the daughters' genes come from the mother: a daughter will have a 50% chance of sharing one of her mother's genes with a sister. Overall, sisters will therefore share around 75% of their genes. In contrast, brothers have a degree of relatedness of 1/2 with their siblings, as they will have a 50% chance of sharing a particular gene derived from their mother. Sisters are therefore more closely related to each other than in animals lacking this form of sex determination, and sisters may actually pass on more of their genes by helping to raise female siblings (with a degree of relatedness of 3/4) than by raising their own offspring (with a degree of relatedness of 1/2). In Hymenoptera colonies, the workers that devote themselves to foraging, nest building, defence, or brood-rearing are generally all females.

Hamilton coined the term *inclusive fitness* to capture the idea that the reproductive success of an individual depends not only on how many offspring it has, but also on the extra fitness it can gain by helping relatives. Later, the term *kin selection* came into use to describe selection that takes account of other relatives as well as immediate descendants. Kin selection is not confined to the Hymenoptera and can be generally applied to any situation in which an individual behaves in apparently altruistic ways towards closely related kin to enhance their inclusive fitness.

As one of the world's greatest experts on the social insects, Wilson was among the first to realize the significance of Hamilton's papers. He became an enthusiastic champion of Hamilton's work and, together with Dawkins, takes great credit for bringing the idea of kin selection to prominence. In *Sociobiology*, Wilson showed how kin selection could be used to explain why many primates, social carnivores, cooperatively breeding birds, and aphids also help their mothers to raise their sisters, as well as why many mammals and birds place themselves at risk by giving warning calls about predators. Not surprisingly, when Wilson turned his thoughts to humanity, he attempted to explain why a group of human beings should apparently be prepared to forgo direct reproductive opportunities, namely the homosexual community, in terms of kin selection. Developing an idea proposed by Trivers, he speculated that:

> the homosexual members of primitive societies may have functioned as helpers … Freed from the special obligations of parental duties, they could have operated with special efficiency in assisting close relatives. Genes favoring homosexuality could then be sustained at a high equilibrium level by kin selection alone (1975, p. 555)

This hypothesis is typical of the final chapter of *Sociobiology*: bold, speculative, and naively insensitive to political connotations. Nonetheless, kin selection has since been invoked to explain a great deal of altruistic behaviour in humans, with some success.

Conflict between parents and offspring

The groundbreaking works of Hamilton were followed by equally influential papers by Robert Trivers at Harvard University. Despite suffering from painful spells of schizophrenia which impaired his normal functioning (Segerstråle, 2000), in a few fertile years in the early 1970s, Trivers single-handedly devised a wealth of sociobiological theory. Like Hamilton, Trivers revealed his genius while still a graduate student, in his case in the same Harvard department as Wilson. He was described by Wilson as a manic depressive of dazzling intellect who periodically would burst into his office and let loose a flood of ideas, some wild and some brilliant. Wilson likened a conversation with Trivers to 'taking a mind-altering and possibly dangerous drug' (1994, p. 325), and confessed that two or three hours with Trivers left him exhausted for the day. Trivers's extraordinary contribution stemmed from his mastery of the gene's-eye view perspective.

Trivers contended that differences in degrees of relatedness would result in conflicts of interest between individuals. In two pioneering papers, 'Parental investment and sexual selection' (1972) and 'Parent–offspring conflict' (1974), Trivers reasoned that, in diploid species, parents should favour equal investment in all of their offspring if the costs of production are equal, while offspring should favour increased investment in themselves rather than in current or future siblings. This is because parents are equally related to all of their offspring, but offspring have greater interest in themselves than in their siblings. Trivers suggested that natural selection would favour traits in offspring that helped them to get as much food and support as possible from their parents before being forced to become self-sufficient, while selection would favour parental behaviour that strikes a balance between their investment in current offspring and saving some energy and resources for the next litter, clutch, or child. Trivers noted how, in many birds, the chicks will vociferously and energetically beg their parents for food around the time of fledging. He also reported how in langurs, baboons, and rhesus monkeys an apparent conflict over weaning often lasts for several weeks. The infant utters a series of piercing cries in its effort to beg milk from its mother, or hitch a ride on its back, while the mother frequently pushes the head away from the nipple or strips the infant away from her body. Trivers interpreted the 'temper tantrums' of young birds and monkeys as an attempt by offspring to manipulate the parent into prolonging the period of parental investment. Prior views of parent–offspring squabbles had treated them as either a non-adaptive consequence of the rupture of the parent–offspring bond or a device promoting the independence of the shy young animal. In contrast, Trivers interpreted it as the outcome of natural selection operating in opposite directions on the two generations.

While the exact dynamics of parent–offspring conflict have proved to be less easily investigated (Bateson, 1994), Trivers's ideas provided huge impetus for further work by biologists and led to a fresh interpretation of parent–offspring interactions in humans. Could the temper tantrums of young children be an attempt to manipulate their mothers into prolonging breast feeding and other forms of parental investment? It turns out that tantrums in 2-year-old children seem to be concerned with the process of establishing autonomy from the parents, rather than with conflicts of interest over weaning (Bateson, 1994). Nonetheless, Trivers's insights stimulated a battery of studies investigating parent–offspring conflict in primates, including humans, and provided the impetus for important advances.

In 1973, Trivers published a paper together with Dan Willard in which they suggested that parents may be selected to invest different amounts of resources in sons compared to daughters, if that would maximize the number of grandchildren they produced. This idea was used by anthropologist Mildred Dickemann (1979) to investigate why some human parents prefer sons or daughters. In many parts of the world, girls are more likely to be killed, abandoned, or deprived of food or medicine than are boys. For example, in China, for every 100 daughters reported to have been born, around 114 sons are registered (compared to 105 in most Westernized countries), with the majority of these missing females having been aborted during pregnancy or killed after birth (Clarke, 2000). Using examples from India, China, and medieval Western Europe, Dickemann suggested that a preference for sons is related to the pattern of wealth inheritance. For example, in early 19th-century India, where all individuals were born into a closely defined socioeconomic class (or caste), sons inherited the wealth of the family while daughters were expected to marry into a higher social strata. The British colonizers of India were puzzled by the lack of daughters in the very highest ranking families, those of the Rajput subcaste, until they realized that most daughters in this caste were being killed soon after birth. The daughters could not be married off and consequently would detract from the abundant wealth that would otherwise be inherited by the sons. These sons were able to take several wives and were therefore of much greater value to their parents than were the daughters.

Dickemann found that female infanticide is more common among high- than low-caste families, allowing her to explain why human sex ratios vary with socioeconomic status in these countries by using Trivers and Willard's theory. However, Dickemann's prediction that families of low socioeconomic status would favour daughters over sons was not upheld by the data, possibly because daughters entail the costs of a dowry, a sum of money or other currency that is required in order to be accepted as a wife. Indeed, testing Trivers and Willard's hypothesis has proven to be more complicated than first expected in humans and other animals (Sieff, 1990; Trillmich, 1996; Brown, 2001). Nonetheless, the application of Trivers and Willard's idea to human beings provided a stimulating new perspective on the question of why parents may treat sons and daughters differently.

Reciprocal altruism

In a paper published in 1971, Trivers introduced the key idea of reciprocal altruism. He suggested that, if unrelated individuals interacted over an extended period of time, an altruistic behaviour which was initially costly to the actor but beneficial to the recipient could be selected if there was a high probability that the altruistic act would be reciprocated between the two individuals on a future occasion. Over time, both individuals would gain more than if they had not cooperated at all. However, the difficulty that must be overcome is the tendency for individuals to cheat and not to reciprocate. Reciprocal altruism may therefore be predicted to occur more frequently in cases where individuals interact with each other on a regular basis and maintain a memory of previous interactions, such that cheating individuals would not receive altruistic benefits in the future. Nonetheless, more subtle forms of cheating, such as never reciprocating quite as much as one receives, might be expected if such individuals can get away with it.

One of the best-known examples of reciprocal altruism occurring in nature has been a study on vampire bats published in 1984 by Gerald Wilkinson of the University of Maryland. After a night of foraging, some individual vampire bats returned to their hollow tree roosts hungry and low in body weight, having failed to find a source of blood. Wilkinson observed other group members in the tree regurgitating food to these individuals, who otherwise risked starvation. As the bats were found to live in relatively stable groups, and returned to the same roost site each morning,

this food exchange may have involved an element of reciprocity. Here the cost of blood donation is small but it can
make the difference between life and death to the recipient,
who is sure to get the opportunity to return the kindness.
While similar evidence for reciprocal altruism has been documented in birds, monkeys, and apes, such data are often
not conclusive and, in most reported cases, alternative
explanations, such as kin selection or mutualisms in which
both individuals obtain immediate benefits, have sufficed.

In fact, human beings may be the animal in which reciprocal altruism most commonly occurs. Trivers argued that
reciprocal altruism is likely to have evolved in the small
stable social groups inhabited by our ancestors over the last
few million years. The system that evolved should allow
humans to reap the benefits of altruistic exchanges, to protect themselves from gross and subtle forms of cheating, but
to practise forms of cheating where profitable. Moreover, he
suggested that selection for reciprocal altruism provides an
explanation for certain characteristics of humans. For
instance, the need for *friendship* is adaptive because it moti-
vates us to find and associate with individuals with whom
we can trade altruistic acts. *Moralistic aggression*, on the
other hand, has evolved so that cheaters will not go un-
punished, while *gratitude* on the part of the recipient of a
kindness is adaptive because it makes the donor believe that
the beneficiary is likely to reciprocate on a future occasion.
Finally, in complex social systems that practise reciprocal
exchange a *sense of justice* is needed as a standard with
which to judge the behaviour of others. While it is possible
to construct alternative explanations for these traits in
humans, Trivers' explanations were both intuitive and
compelling. As a result of Trivers' work, economists have
become particularly interested in whether humans act

reciprocally when they bargain over the distribution of money or resources.

Evolutionary game theory

In turn, ideas from the study of economics have been influential within evolutionary theory. Evolutionary game theory is a way of thinking about evolution when the advantage of behaving in a particular manner depends on what other individuals are doing. Building on earlier game theory ideas of economists, Maynard Smith and Price (1973) pioneered this evolutionary approach. The goal of the exercise is to try to work out which behaviour is the most stable strategy, on the assumption that over millions of years of evolution this is what would have evolved. For instance, in deciding whether to engage in a fight over a resource, an individual may adopt the strategy of 'always attack', 'never start a fight', or perhaps 'always attack when challenged'. Other strategies are conditional, for example, 'attack only if the opponent is smaller', or 'retreat only if the opponent is larger'. If all of the possible strategies are pitted against each other, for example on a computer or by constructing a mathematical model, the winning strategy can be determined. This strategy is known as the *evolutionarily stable strategy* (ESS), and if it is adopted by all members of the population no other strategy could replace it.

Evolutionary game theory was originally applied to the study of animal conflicts over resources, notably by Geoff Parker, but has since been successfully used to investigate how individuals might behave in a wide variety of situations, including whether to forage or steal food, when to cooperate with another individual, and what information to share with others. In each case, the conclusion that a particular ESS should be reached can be tested by investigating whether something like the ESS is observed in natural populations of animals. There is no doubt that the quantitative rigour that the ESS framework imposes on thinking in animal behaviour has made it an indispensable tool for the study of adaptation.

In *The Selfish Gene* (1976), Dawkins made the ESS a cornerstone of his argument. For instance, he used evolutionary game theory to illustrate the conflicts between the sexes that may arise over parental investment. Developing ideas introduced by Trivers in 1972, Dawkins points out that, while both parents may want offspring, for each there may be some advantage to investing less than their fair share of time and resources in the child. This is because, if they can manipulate their partner into bearing the bulk of the costs of a successful rearing, they have still effectively passed on their genes but have extra time and resources to devote to further reproduction. Trivers had pointed out that when species are classified according to the relative parental investment of the sexes in their young, in the vast majority of vertebrate species the male's only contribution is his sex cells, which the female's contribution clearly exceeds by a large amount.

Dawkins suggested that one strategy that a female might adopt to get round this problem (labelled 'coy') is to seek out reliable males and subject them to an extended courtship to assess their fidelity. In fact, Dawkins proposes two male strategies (faithful and philanderer) and two female (coy and fast). Coy females will not copulate until after a long courtship, while fast females will copulate immediately. Faithful males are prepared to undergo the extended courtship and will help rear the young. In contrast, philanderers are not interested in courtships, will leave if they do not copulate immediately, and will not help raise young. For illustrative purposes, Dawkins allocates arbitrary values for the various costs and benefits: 15 units for each successful child raised, 20 for the cost of rearing, and 3 for the cost of courtship. With these particular values a mixed ESS is reached in which 5/6 of the females are coy and 5/8 of the males are faithful, or when each female is coy 5/6 of the time, and each male faithful 5/8 of the time. Thus in this particular instance, males are more likely to be promiscuous than females are to be fast.

At the end of this discussion, Dawkins reflects on the extent to which this reasoning applies to humans. Although the values for the costs and benefits and the choice of strategies are openly arbitrary and different values would yield different results, and despite acknowledging that in humans promiscuity is probably more affected by culture than by evolved dispositions, Dawkins concluded 'it is still possible that human males in general have a tendency towards promiscuity' (1976, p. 164). Evolutionary game theory has been less commonly applied to life history strategies of human beings than those of other animals. However, this methodology still remains important for those interested in how human beings make decisions between choices when the outcomes of those choices are influenced by the decisions of others, and it is beginning to make an impact in evolutionary economics.[2]

---

[2] This development can be seen in some recent contributions to the journal *Games and Economic Behavior*, and in the books of Fernando Vega-Redondo (1996) and Ken Binmore (1998).

The human sociobiology debate

The revolutionary ideas of Williams, Hamilton, Trivers, and Maynard Smith were of huge importance for the study of animal behaviour. The controversial aspect was the application of sociobiology to human beings and Wilson, in particular, received much of the attention in this regard. Wilson later summarized his ideas on human sociobiology as follows:

> Human beings inherit a propensity to acquire behavior and social structures, a propensity that is shared by enough people to be called human nature. The defining traits include division of labor between the sexes, bonding between kin, incest avoidance, other forms of ethical behavior, suspicion of strangers, tribalism, dominance orders within groups, male dominance over-all, and territorial aggression over limiting resources. Although people have free will and the choice to turn in many directions, the channels of their psychological development are nevertheless ... cut more deeply by the genes in certain directions than in others. While cultures vary greatly, they inevitably converge toward these traits. (1994, pp. 332–3)

These views provoked strong opposition, and before long the dispute had become a media event. Almost immediately, a vocal countermovement of hostile critics of human sociobiology sprung forth. Anthropologists, psychologists, sociologists, and some prominent biologists bitterly repudiated the sociobiologists' findings, lambasted their methods, and charged them with prejudicial storytelling. The history of this controversy has itself resulted in numerous papers and books (e.g. Segerstråle, 1986; 2000). Here we will provide only a short account of the political opposition to human sociobiology and spend more time discussing the scientific arguments.

The immediate opposition that arose against the ideas in the final chapter of *Sociobiology* were from politically active academics. A Boston-based collective of scientists and social scientists came together to form the Sociobiology Study Group, which soon affiliated itself with Science for the People, a nationwide organization of activists begun in the 1960s to expose the misdeeds of scientists. At a time when student demonstrations against the Vietnam war were commonplace, and following on from a recent melee over race and intelligence testing, motivating students to protest against allegedly subversive or dangerous scientists did not prove to be difficult. The Sociobiology Study Group was dominated by Marxist and left-wing scholars from Harvard. Two of the most prominent and vocal were evolutionary biologists Richard Lewontin and Stephen Jay Gould, both of whom worked in the same building as Wilson at Harvard. In fact, the group of scientists and social scientists most openly critical of human sociobiology met together in Lewontin's office, directly below that of Wilson.

In a letter published in the *New York Review of Books* on 13 November 1975 the Sociobiology Study Group declared that human sociobiology was not only unsupported but tended to provide a genetic justification of the status quo, and perpetuated inequalities on the basis of sex, class, and race (Allen *et al.*, 1975). They accused sociobiology of being 'reductionist', 'biologically determinist', and motivated by ignorance and chauvinism, took issue with the hypothesis that society reflected biological imperatives and linked sociobiology to former disturbing applications of evolutionary theories:

> These theories provided an important basis for the enactment of sterilization laws and restrictive immigration laws by the United States between 1910 and 1930 and also

*the anti Semitism / matter
World Soc. word not a
of eugenics.*

for the (eugenics) policies which led to the establishment of
gas chambers in Nazi Germany. The latest attempt to
reinvigorate these tired theories comes with the alleged
creation of a new discipline, sociobiology (Allen *et al.*, 1975).

Wilson clearly believed that good scientists should have
the courage to address difficult issues and to keep going if
the flak starts to fly. Convinced that what he had said in
Sociobiology was justifiable, Wilson went on the offensive.
He castigated his critics as political extremists who perpetuated the myth of the mind being a blank slate at birth
(*tabula rasa*) only because it was consistent with their naive
dream of a perfect society. At the same time, he extended
his research into human behaviour and in 1978 published
On Human Nature, an immediate bestseller that won a
Pulitzer prize. Wilson continued to make bold claims that
were to keep him at the heart of the controversy, suggesting
for example that differences in the behaviour of men and
women reflected past evolutionary events and could only be
eradicated at some cost to society. The debate became
highly charged and politicized and in 1978 feelings ran so
high that at one major scientific meeting a group of demon-
strators took over the stage when Wilson was about to
speak, shouting out that Wilson was a racist, and then
dumped a pitcher of ice water on his head.

One critic that Wilson could not dismiss as biologically
naive was his colleague Richard Lewontin. The conflict
between these two members of the same Harvard department, each an outstanding scientist and an authority on
evolutionary biology, born in the same year, is one of the
most beguiling features of the sociobiology controversy.
Ironically, it was largely through Wilson's efforts that
Lewontin was recruited to Harvard in the early 1970s, only
for Lewontin seemingly to bite the hand that had fed him.

However, Wilson later acknowledged Lewontin as a worthy adversary and suggested that without Lewontin the controversy would not have been so intense or attracted so much attention (Wilson, 1994). Lewontin gave countless public lectures criticizing sociobiology and wrote endless hostile reviews of sociobiological books. There was no doubting Lewontin's credentials—he was one of the world's most brilliant population geneticists and Steve Gould once described him as the cleverest scientist he had ever known. However, Lewontin has always been very open about his strong political views and he is a man of great integrity. At a very early age Lewontin had been elected to the National Academy of Sciences, only to resign in protest over the Academy's sponsorship of military research.

In the cauldron atmosphere of the sociobiology debate it was easy to take sides and dismiss Wilson as truly motivated by prejudice or Lewontin by his Marxist ideology. In reality their differences are principally neither related to politics nor prejudice, but over science (Ruse, 1999; Segerstråle, 2000). Wilson was the kind of scientist who relished the challenge of major problems, saw the big picture, and constantly wanted to push fields forward by developing and synthesizing new theory. In contrast, Lewontin was much more cautious, suspicious of sweeping statements and unsupported speculation, and deeply sensitive to how vulnerable biological arguments are to abuse. For Lewontin, science had to be as correct as possible because mistaken scientific theories lent themselves to political abuse (Segerstråle, 2000), a belief which, as we saw in the previous chapter, can find justification in the history of using evolution to interpret humanity.

What most of the critics saw as an insurmountable problem for human sociobiology was the special status of the human species, based largely on our capacity for language and culture (Segerstråle, 2000). To Wilson's credit, he was prepared to take on board some of the criticism levelled at him and other sociobiologists. In his autobiography he states that 'it was obvious to me that human sociobiology would remain in trouble, both intellectually and politically, until it incorporated culture into its analyses' (1994, p. 350). In 1979, Wilson was joined by a postdoctoral researcher called Charles Lumsden, a Canadian theoretical physicist. They decided to develop mathematical models that explored the relationship between genes and culture, and within two years had published a book entitled *Genes, Mind and Culture.* In this book, Wilson accepted that human culture has shared, socially transmitted features that make it distinct from other aspects of the human pheno-type. In many respects, this book is a striking departure from the verbal accounts of human nature in *Sociobiology.* It was an attempt to put human sociobiology on the firm theoretical, quantitative basis that the critics had found lacking in Wilson's earlier work. Lumsden and Wilson endeavoured to go back to basics, to absorb as much as they could about human psychology, anthropology, and social behaviour, and to construct a new body of theory that could address questions at the heart of the sociobiology debate. They reasoned that human behaviour is influenced by culture, elements of which, called 'culturgens', are trans-mitted between individuals (for example, particular ideas, beliefs, or patterns of behaviour) and that society could be depicted as the collective distribution of culturgens in the population. They suggested that whether an individual adopts a particular culturgen depends on the characteristics of his or her brain, which is subject to genetic biases via developmental processes called 'epigenetic rules'. Thus even though individuals learn aspects of their culture, they are programmed to acquire some culturgens more easily than others, as natural selection has favoured individuals with epigenetic rules that bias them towards adaptive behaviour. As a result of their mathematical treatise, they reach a number of important conclusions, for example, that *tabula rasa* is an unlikely condition for the human mind, that culture can affect the rate of genetic evolution, and that it takes approximately a thousand years for genetic predispositions that bias culture to evolve.[3]

Wilson (1994) confesses to being puzzled by the fact that this work was largely ignored. However, published as it was in the midst of the sociobiology debate, with Wilson well established as a pariah to the social sciences, this book was never going to be embraced or even judged objectively. When one also considers its highly technical nature and the fact that Lumsden's mathematical methods would have been unfamiliar to virtually all readers, *Genes, Mind and Culture* never stood a chance. For almost everyone, this theory was completely opaque and, as a consequence, their assessment of it was heavily influenced by hostile reviews (see, for example Maynard Smith and Warren, 1982; Kitcher, 1985). In our judgement this is a shame, as Wilson had made a genuine attempt to respond positively to his critics. Unfortunately, by then it was too late—sociobiology had become a dirty word to many social scientists and most of them were highly suspicious of Wilson. He had failed to bring to fruition his vision of another new synthesis, that of

---

[3]   In their review of *Genes, Mind and Culture*, Maynard Smith and Warren (1982) criticized the conclusion of 'a thousand year rule', and asserted that it followed from the authors' assumptions of strong selection, weak culture, and high heritability.

biology and the human sciences. For all his credibility in biological circles, Wilson's human sociobiology had been resoundingly rejected by the very people he believed could most benefit from it, the social scientists.[4]

Critical evaluation

Despite its turbulent beginnings, has sociobiology increased our understanding of social behaviour, or were the social scientists right to dismiss its methodologies and ideas? Perhaps the first point to make here is that the ideas provided by Hamilton, Trivers, Williams, and Maynard Smith, which were presented by Wilson in the first 26 chapters of *Sociobiology* and by Dawkins in *The Selfish Gene*, have revolutionized the study of animal social behaviour, helping to dismiss naive group selectionism and providing explanations for the behaviour of diverse animal species. The work carried out using these ideas continues under the name of behavioural ecology. Virtually all of the controversy has surrounded the application of these ideas to human behaviour. We will review the charges of reductionism and genetic determinism, prejudice and story-telling, and then return to consider the rejection of sociobiology by the social scientists.

---

[4] The extent to which human sociobiology is regarded as having been rejected by the social science community depends in part on definition and perspective. If human sociobiology is defined loosely to include all current evolutionary approaches to the study of human behaviour then it could be regarded as thriving. For instance, there are several peer-reviewed sociobiological journals, the Human Behaviour and Evolution Society has consistently increased in numbers, and there have been more than 200 published books on this topic since 1975, almost all supportive (Wilson, 2000). However, it remains the case that the vast majority of social scientists are unsympathetic to an evolutionary perspective

Reductionism and genetic determinism

Human sociobiology was most frequently denounced for the sins of reductionism and genetic determinism. Critics angrily chastised sociobiologists for suggesting that the behaviour of individual human beings is determined by genes, and that complex behaviour could be reduced to genetic effects. Again, Lewontin was a strong opponent to this view of behavioural development and to the idea of reductionism. For instance, in *Not In Our Genes* (1984), Rose, Lewontin and Kamin wrote:

> Sociobiology is a reductionist, biological determinist explanation of human existence. Its adherents claim…that the details of present and past social arrangements are the inevitable manifestation of the specific action of genes.
> (p. 236)

At first sight this criticism appears unfounded. Wilson had been keen to point out that he was not attempting to show that human behaviour patterns were solely influenced by genes. The behaviour of all animals is a product of the interaction between genes and environment, while learning and experience allow individuals to acquire novel information. In other words, genes exert a diffuse influence on human activities and the arguments over genetic determinism boil down to just how much influence different people think genes have. All of the major protagonists of sociobiology have bent over backwards to be clear that they do not believe in genetic determinism. Yet much confusion still occurs today when the term 'a gene for' a behaviour is used, and when the reproductive success of individuals is said to be measured in terms of the number of genes passed on. The possibility that there may be a genetic influence on behaviour, which may be inherited by the next generation, does not imply that such behaviour patterns are solely determined by one or more genes or that such behaviour patterns are fixed and inevitable. Rather, it is based on the assumption that there is a multitude of non-genetic influences on behavioural development but, as such influences are typically not heritable, they can be ignored in evolutionary analyses. Nonetheless, many biologists have expressed the concern that the 'gene for' language might lead to a neglect or trivialization of developmental processes (e.g. Bateson, 1981). In reality, much of the debate was not about genetic determinism at all, but rather genetic constraints and propensities. Wilson suggested that 'Rather than specify a single trait, human genes prescribe the *capacity* to develop a certain array of traits' (1978, p. 56). Whether an individual expresses a particular behaviour pattern may depend upon a myriad of factors encountered by that individual over a lifetime, including social and cultural influences.

In contrast to Lewontin, Wilson did believe that a reductionist approach to the study of behaviour was appropriate. One dictionary definition of reductionism is 'the belief that complex data and phenomena can be explained in terms of something simpler'. This method of understanding the world is applied throughout science, and it would seem more of a virtue than a sin. Wilson acknowledged that this is the way that he thinks, and sees nothing wrong with it. However, Lewontin appeared to have something else in mind:

> By reductionism, we mean the belief that the world is broken up into tiny bits and pieces, each of which has its own properties and which combine together to make larger things. The individual makes society, for example, and society is nothing but the manifestation of the properties of individual human beings. (Lewontin, 1991, p. 107)

Lewontin argued that there are properties of society and of social institutions that do not reduce to properties of people and that, sometimes, these cannot be ignored. In a sense, Wilson acknowledges this in *Genes, Mind and Culture* with his treatment of culture as a dynamic process in its own right. However, Wilson clearly regarded culture as being constrained by our biological heritage, as exemplified by his famous phrase that 'the genes hold culture on a leash' (1978, p. 172). The majority of social scientists are still content to assume that culture is responsible for most human behaviour and that the role of genes is small enough to be of little relevance to the study of humans. There is nothing reductionist or deterministic about the challenging of this assumption, and Wilson is surely entirely justified in asking whether there might be adaptive biological influences that prevail despite the influences of culture.

[handwritten margin note: n: "culture is a product of biology"]

Prejudice

Some of the statements Wilson makes in *Sociobiology* left him wide open to accusations of sex, class, and race prejudice. Critics and historians alike have concluded that Wilson's writings on humanity reflect the values of his upbringing in the American south and his undergraduate years at the racially segregated University of Alabama. What is apparent is that, in *Sociobiology*, Wilson does make some injudicious statements and that these provoked much hostility. For instance, his views on sex differences sometimes appeared to promote the status quo, while his views on possible differences in mental aptitudes between races were easy targets for attack. Another example is his discussion of whether genes that affect success and an upward shift in status 'would be rapidly concentrated in the uppermost socioeconomic classes' (p. 554). In fact, the next page reveals that Wilson is quite aware of at least some of the counterarguments; for example, that society is too fluid and gene flow too extensive to maintain such genetic class differences. Yet, while playing the intellectual game of devil's advocate, Wilson seemed prepared to entertain perspectives that would be anathema to more politically astute scientists and had apparently spared little thought for the repercussions of his deliberations. Wilson claimed to have been ignorant of the possibility of outrage at his work. Maynard Smith stated he had disliked the last chapter of *Sociobiology* and later remarked 'It was absolutely obvious to me—I cannot believe Wilson didn't know—that this was going to provoke great hostility' (Segerstråle, 1986). However, in his autobiography, Wilson reiterated his political naivity. Nonetheless, he did recognize that:

> Mine was an exceptionally strong hereditarian position for
> the 1970s. It helped to revive the long-standing nature–
> nurture debate at a time when nurture had seemingly won.
> The social sciences were being built upon that victory.
> (1994, p. 333)

Many critics of sociobiology were concerned that, irrespective of whether sociobiologists themselves harboured prejudices, sociobiological arguments were vulnerable to racist and prejudicial interpretations (Rose *et al.*, 1984). When, in 1981, biologist Steven Rose wrote a letter to the journal *Nature* revealing that an extreme right-wing organization had been using sociobiological writings to support their racist creed, for many these fears appeared justified (Segerstråle, 2000). Rose challenged leading sociobiologists to disassociate themselves from these neo-Nazi views and Maynard Smith, Dawkins, and Wilson all immediately wrote firm replies stating that there could be no justification for racism in sociobiology (Segerstråle, 2000).

The sociobiological claim that human social organization reflects a history of natural selection led advocates and critics alike to the conclusion that the current state of society is in some sense optimal. Wilson also warned that humans can manufacture an equal society only at a cost and asserted that our genetic heritage may render it imposs-   ⊢ ible to mould society in certain directions. For those who regarded American society in the 1970s as riddled with race, class, and gender prejudices, this was an unsavoury message. However, Wilson's view was not shared by all human sociobiologists. In his *Darwinism and Human Affairs* (1979), University of Michigan biologist Richard Alexander stated that an evolution- ary interpretation of human history does not imply a deterministic future and that conscious awareness of our biology allows us to release the bind to our history   *N., It wn't.* of fitness maximization. Moreover, Richard Dawkins, in his response to the neo-Nazi article exposed by Rose, stressed that genetically inherited traits were far from unmodifiable:

> What is really wrong with the National Front quotation is not the suggestion that natural selection favoured the evolution of a tendency to be selfish and even racist. What I object to is the suggestion that if such tendencies had evolved they would be *inevitable* and *ineradicable*; the suggestion that we are stuck with our biological nature and can't change it (Dawkins, 1981; italics added).

Dawkins went on to charge critics of sociobiology with propagating the 'myth' that sociobiologists believed in the inevitability of genetic effects.

Story-telling

Perhaps the most telling criticism of sociobiology is that many of the hypotheses were no more than plausible stories for the origin of human behavioural traits. For example, Rose *et al.* (1984) complained that:

> imaginative stories have been told for ethics, religion, male domination, aggression, artistic ability, etc. All one need do is predicate a genetically determined contrast in the past and then use some imagination, in a Darwinian version of Kipling's *Just So Stories*. (p. 258)

Ironically, the problem stems from the fertile nature of evolutionary reasoning. Inventing evolutionary stories is a seductively easy exercise. If we were to attempt to explain human sex differences in average height, for example, we could come up with numerous evolutionary hypotheses. For instance, on average, men may be taller than women because, in the past, females preferred to mate with tall males or perhaps extra height gave tall males an advantage while hunting in the savannah, while searching for prey or while throwing spears, or gave some advantage during fights with other males. Perhaps extra shortness gave females an advantage while gathering, making it easier to collect plant material on the ground, or leaving them less visible to predators. Height has the advantage that it is a characteristic that is manifest in the fossil record, that it is easily quantified in contemporary populations, and that the sex difference is unlikely to result from social or economic variation. Behavioural and psychological attributes such as promiscuity or intelligence are not even subject to these constraints. Developing hypotheses is a fundamental part of the scientific process and part of the enduring appeal of evolutionary theory is that it is such an effective instrument for doing so. However, this only makes it even more important that the hypotheses should not only be potentially testable but actually tested.

Let us return to Wilson's kin selection explanation for homosexuality. Rose *et al.* (1984) point out the weaknesses in this argument. First, there is no evidence that in the past homosexuals had fewer offspring than heterosexuals. Secondly, there is no satisfactory evidence that homosexuality has a genetic basis (see our discussion of heritable traits in Chapter 7). Thirdly, there is no evidence that homosexuals, either now or in our evolutionary past, helped their relatives to raise offspring any more than heterosexuals. Given the politically sensitive nature of the topic, it is easy to see how the superficiality of this explanation could be regarded as irresponsible.

The direct comparison between the behaviour of human beings and other animals had been a method commonly used by the popularizers of human ethology. Wilson had tried to introduce a little rigour into the comparative analysis of human behaviour. He clearly believed that he could do better than predecessors such as Morris, Ardrey, and Tiger and Fox, whose 'particular handling of the problem tended to be inefficient and misleading' (Wilson, 1975, p. 551). Wilson suggested that traits that vary from species to species or genus to genus are so evolutionarily unstable that it is foolhardy to use them to make comparative inferences about humans. Only those characters that are constant at the level of family, or order, may be sufficiently stable to have persisted in relatively unaltered form during the evolution of modern humans. Such conservative traits, Wilson suggested, might warrant an evolutionary explanation. While this careful approach is admirable, much of the comparative work was based on views of primate social behaviour and hunter–gatherer lifestyles that are now interpreted differently. That is not to say that comparative methods, which have come a long way since 1975, cannot be employed to draw testable inferences about human behaviour (see Harvey and Pagel, 1991). Even so, because all of our close relatives are extinct and because we know so little about other species in the *Homo* genus, or the family Hominidae, it is difficult to be sure that behavioural traits that appear to be homologous with apes (that is, inherited from common ancestors) really are so.

The works of Sarah Blaffer Hrdy provide examples of how taking a careful comparative approach can help us to understand human behaviour and can provide evidence that dispels outdated views on human nature. Hrdy was an undergraduate and postgraduate student in Anthropology at Harvard in the pre- and post-sociobiology years, and her mentors included Wilson and Trivers, as well as the primatologist Irven DeVore. During an undergraduate lecture attended by Hrdy, DeVore mentioned a report by Japanese primatologists working in India that described adult male langur monkeys grabbing infants from their mothers and biting them to death. This behaviour was assumed to be 'pathological' and caused by high population densities. Hrdy was intrigued and carried out her postgraduate work on these langurs. Her studies in the field showed that male attacks on infants only occurred when new males entered the breeding group. Using Trivers' viewpoint, Hrdy (1977) suggested that males would be selected to eliminate unweaned infants that were not their own, as females would then ovulate sooner than if the infant had lived and continued to suckle. The willingness of female langur monkeys to mate with infanticidal males was viewed by Hrdy as an adaptive strategy on the part of females in response to the high turnover of males in the group. The mother of an infant would make some attempts to prevent such attacks, yet would often mate with the male that had just killed her offspring.

In 1981, Hrdy published *The Woman that Never Evolved*, in which she put forward her views on the evolution of women and other female primates. These ideas followed a tradition of other women, including Antoinette Brown Blackwell (1875) and Clémence Royer (1870), who had responded to the publication of Darwin's *The Origin of the Species* by pointing out his lack of emphasis on the evolution of female behaviour. However, not until the end of the 20th century was the behaviour of female primates to receive the attention that it deserved. In her book, Hrdy pointed out that the view of women and other female primates as sexually and socially passive animals was just not backed up by the available evidence. Hrdy demonstrated how female primates have strategies of their own and that the social relations of female primates have a great influence on the dynamics of social groups. In her latest book, *Mother Nature: Natural Selection and the Female of the Species* (1999), Hrdy argues that infanticide by invading males may be less common in humans than in other primates, such as langurs, gorillas, and chimpanzees, but suggests that the potential threat from members of our own species may partially explain why human infants show a fear of strangers. Hrdy's work shows how the application of thoughts on animal behaviour can be applied to our understanding of human behaviour, and highlights the fact that sociobiology can result in the dispelling, rather than the bolstering, of prejudicial views on human nature.

The rejection of sociobiology by social scientists

For most social scientists, the real problem with socio-biology was not that it was reductionist, or deterministic, or prejudicial, nor that sociobiologists were encroaching on their territory. Rather, it was that too much human socio-biology was dilettante. In their enthusiasm, human socio-biologists capriciously flitted from one topic to the next, often concocting superficial stories without ever stopping to develop a solid understanding of the topic or read the social science literature. The work was frequently carried out with a religious fervour for serving up biological home truths, with scarcely a thought for alternative non-evolutionary explanations. In fact, the attacks on their work from the social scientists may even have caused the early human sociobiologists to close ranks and to avoid cri-ticizing each other's work. Had sociobiologists been more responsible in their application of evolutionary theories of human behaviour, for instance, by asking whether they had evidence for their suppositions, con-sidering the merits of non-evolutionary explanations, and utilizing the data and insights collected by social scientists, they might have been much less likely to provoke a negative reaction.

The tragedy is that many of the good ideas generated by sociobiology were dismissed because of its failings. In his autobiography, Wilson complains that the sociological or cultural model of his critics is assumed true unless proven false beyond any possible doubt, while biological hypo-theses are assumed to be false unless evidence is completely unassailable in their support. While Wilson clearly has a case, there is a strong argument that this bias is justified. Perhaps behavioural science should have as a starting point, or null hypothesis, the assumption that all types of society are possible and that no behavioural differences between subsections of the population are impossible or onerous to eradicate. If a higher standard of science were maintained by those taking an evolutionary approach to human behaviour, evolutionary explanations might be less open to abuse.

Sociobiology has undoubtedly contributed to an understanding of human behaviour, particularly with regard to topics such as cooperation, conflicts of interest, parental investment, and female sexual behaviour. Moreover, sociobiology gave us a new set of methods with which to explore human behaviour, including the gene's-eye view, kin selection, evolutionary game theory, and reciprocal altruism. Selfish genery was a major advance in thinking about animal behaviour and evolution, and it applies equally to humans. What should be remembered is that 'selfishness' as a human trait is not implied by the selfish gene view, and cooperative behaviour may equally result from sociobiological reasoning. For example, sociobiological theory has shown us that much altruistic behaviour in humans may be explained in selfish gene terms. Neither does the observation that certain human behaviour patterns or differences have evolved imply that they are 'right', a mistake that sociobiologists have repeatedly cautioned against. In *The Selfish Gene*, Dawkins wrote:

> I am saying how things have evolved. I am not saying how we humans morally ought to behave (1976),

while Wilson stated:

> There is a dangerous trap in sociobiology, one which can be avoided only by constant vigilance. The trap is the naturalistic fallacy of ethics, which uncritically concludes that what is, should be (1975b).

A comparison of human beings with other social animals—particularly primates—can reveal what is unique to the human species and what is similar to other animals (Hrdy, 1999; Brown, 2000). Moreover, a rigorous analysis of animal behaviour across a broad range of species allows the abstraction of general principles that may also apply to human beings. In addition, the detailed study of animal behaviour and careful use of similar methodologies with human beings may help us to understand whether current views that, for example, particular sex differences in behaviour are natural and deeply rooted in our animal heritage, are based on false understanding of animal social behaviour. Evolutionary biology can dispel prejudicial myths as well as support them.

A new dawn?

In 1977, Wilson received the National Medal of Science from President Carter for his contributions to the new discipline of sociobiology. Wilson has repeatedly been nominated for a Nobel prize for starting the field and his many other contributions to science. Perhaps his unique combination of creativity, courage, and political naivity will eventually be regarded as the catalyst for his dream of an integration of biological and social sciences. While this vision has yet to be fully realized, and most anthropologists and psychologists continue to ignore sociobiological methods, there are signs that related approaches are beginning to emanate that have a greater resonance with social science thinking.

The first strand to emerge was *human behavioural ecology*, mainly carried out by biological anthropologists who have explored to what extent human behaviour is adaptive, under the assumption that much culture is evoked

by various features of the social and ecological environment. Human behavioural ecologists were generally
from an anthropological background, rather than being
biologists that had fleetingly turned their hands to the
human sciences, a shift that in itself led to significant
advances. A few years later, the second strand was to
emerge, now named *evolutionary psychology*, which has
generally been carried out by academic psychologists
searching for the evolved psychological mechanisms that
underpin the universal mental and behavioural characteristics of humanity. In the final chapter of *The Selfish
Gene*, Dawkins presented the idea of the meme, devised as
an alternative to sociobiological explanations of culture,
which has stimulated a large number of scientific and popular writings on cultural evolution, collectively known as
memetics. The field of *gene–culture coevolution* was emerging at around the same time, developing ideas and methods
along the lines of Lumsden and Wilson's *Genes, Mind and
Culture*. However, this work is practised by some of the
theoretical population geneticists and anthropologists that
had been among sociobiology's critics. These four new
fields are described in Chapters 4, 5, 6, and 7.

While human behavioural ecology, evolutionary psychology, memetics, and gene–culture coevolution differ in
important respects from sociobiology, and proponents of
each would regard these differences as a major advance, all
germinated roots in, and owe a debt to, the sociobiological
era. In some ways, human sociobiology could be described
as ongoing, although many researchers avoid using this
term to describe their work as a result of the controversy
that followed Wilson's publications. Other researchers, such
as Hrdy pride themselves in describing their work as human
sociobiology out of respect for the positive contribution

that the application of methods from animal behaviour to the study of human beings has provided. That sociobiology could spawn a wealth of new evolutionary approaches to the study of human behaviour is a testament to the rich, fertile, and pluralistic nature of its theory. For human sociobiology there were to be new dawns.

CHAPTER 4

Human behavioural ecology

While the human sociobiology debate was raging, a number of anthropologists decided to go out and test sociobiological ideas with real data from human populations. They started by asking questions such as 'do human beings exhibit optimal strategies during foraging?' and 'do people alter the number of offspring they raise depending upon their environment?' Their main premise was that human behavioural strategies are adaptive across a broad range of ecological and social conditions.

Considerable attention was already being paid within anthropology to environmental and ecological influences on behaviour but from a group-selectionist point of view (Cronk, 1991). In contrast, early evolutionary or *Darwinian anthropologists* were interested in whether human beings might be able to alter their behaviour flexibly, depending upon present circumstances, to maximize their own reproductive success. Such research is usually now referred to as *human behavioural ecology*. Traditional anthropology places emphasis on the influence that culture has on the behaviour of individuals. Instead, human behavioural ecologists are interested in how an individual's behaviour is influenced by the environment in which he or she lives and how the alternative behavioural strategies that people adopt produce cultural differences.

According to Borgerhoft Mulder:

> The aim of modern human behavioural ecology is to
> determine how ecological and social factors affect
> behavioural variability within and between populations. In
> one sense its hypotheses are viewed as an alternative to the
> more traditional anthropological belief in an unspecified
> force of 'cultural' determination. In another sense,
> behavioural ecological anthropology can be seen as adding
> the study of function to investigations of causation,
> development and historical constraints that were already
> well established in the social sciences. (1991, p. 69)

Early human behavioural ecologists were greatly influenced by the progress being made in the study of animal
behaviour by the use of new theories of optimization and
life history strategies, which linked the behaviour patterns
of individuals to their physical and social environment.
They began studying humans as if they were any other
animal species, observing what individuals actually did in
their lives and comparing this to the predictions made by
their evolutionary hypotheses. As early as 1956, the British
evolutionary biologist J. B. S. Haldane had argued that the
behavioural differences shown by contrasting human
groups were responses to particular environments, and that
these different patterns of behaviour were exhibited by
human beings with basically similar genetic compositions.
This idea was later reiterated by William Irons in the landmark book *Evolutionary Biology and Human Social
Behaviour: an Anthropological Perspective* (1979), which
Napolean Chagnon and he edited. Irons and Chagnon,
both at Northwestern University, were highly influential in
teaching evolutionary approaches to a younger generation
of anthropologists, while at the University of Utah, Kristen
Hawkes, under the guidance of biologist Eric Charnov,

began teaching human behavioural ecology from the late 1970s. The works of zoologists Richard Alexander (1974) and Robert Hinde (1974) provided additional impetus to this emerging field. John Crook, who was a pioneer in the study of animal socioecology and subsequently moved on to work with human beings, also had a major impact by showing that social systems could be seen as ecological adaptations (Crook, 1964, 1965; Crook and Gartlan, 1966).

One of the most influential characters in the establishment of this new discipline was Irven DeVore, an anthropologist working in Harvard at the same time as Wilson. DeVore had been a student of the physical anthropologist Sherwood Washburn at the University of Chicago. Although DeVore was a social anthropology student with no interest in primates, Washburn sent him to Kenya in 1958 to carry out a pioneering study on the social lives of baboons (Kuper, 1994). Washburn was determined that studying non-human primates would provide information about the evolution of human behaviour. He also believed that field studies of contemporary African hunter–gatherers might help us to understand the ways in which early humans had adapted to environmental pressures. DeVore and graduate student Richard Lee were later to initiate an important study of the !Kung Bushmen of the Kalahari.

At Harvard, DeVore was won over by the ideas of Robert Trivers and became a great supporter of Trivers during his difficult early career (Segerstråle, 2000). In comparison, Washburn bitterly rejected sociobiology and continued to maintain his group-selectionist views on human evolution, causing an irreparable rift between the two. Throughout the 1970s, DeVore presided over a thriving research group with a rich atmosphere that resembled an intellectual salon, with distinguished visitors regularly dropping by, and sociobiological ideas and methods constantly being discussed. For a number of years prior to the publication of Wilson's *Sociobiology*, DeVore and Trivers had been teaching a course on using evolutionary biology to investigate human behaviour in the anthropology department at Harvard. Although DeVore and Wilson were friendly colleagues, DeVore was not an avid supporter of Wilson's version of sociobiology as applied to human beings, partly because he felt that Wilson lacked a comprehensive grasp of the anthropological literature. He later stated that:

> When *Sociobiology* was near completion, Ed sent me the last chapter. It was not that I disagreed with him. I wanted him to have written a different book. I felt it was nothing like the whole story. Ed was naïve in many ways in those days. It was not that he had no respect for the social sciences; he had so many other things on his plate. I kept thinking that Trivers and I should have done this! (Segerstråle, 2000, p. 81)

The principal goal of human behavioural ecology is to account for the variation in human behaviour by asking whether models of optimality and fitness-maximization provide good explanations for the differences found between individuals. An overriding assumption is that human beings exhibit an extraordinary flexibility of behaviour, allowing them to behave in an adaptive manner in all kinds of environments. The precise causes of this behaviour are of lesser interest, and many human behavioural ecology researchers avoid detailed discussion of how psychological or cultural factors may influence the expression of particular strategies. The initial work carried out by human behavioural ecologists in the late 1970s and early 1980s focused mainly on ecological questions, particularly foraging behaviour, a topic that was receiving considerable attention by animal behaviourists. For example, data on diet choice among hunter–gatherer populations were compared to the food items available, and models were produced to test whether individuals were foraging optimally, in terms of gaining the greatest possible number of calories per hour of foraging. Subsequently, the field broadened out to address problems concerned with social relationships and conflicts. More recently, human behavioural ecologists have turned their attention to various aspects of the human life history, such as the evolution of menopause, senescence, sex-biased parental investment, and variation in reproduction in response to ecology.[1]

Most of the research carried out within human behavioural ecology has been on small communities in remote regions of the world, often exposed to relatively little contact with Western society. The data are collected by direct observation of behaviour and by use of interview and historical material. Thus the subjects and methods used resemble similar research carried out by non-evolutionary anthropologists, particularly ethnographers, but with a different theoretical and epistemological framework. However, such detailed data collection requires immense effort and time. Good data sets on such communities are only now emerging from a number of research groups, including data on the Ache of Paraguay (studied by Hill, Hurtado, Kaplan, and Hawkes); the Kipsigis of Kenya (studied by Borgerhoff Mulder and colleagues); the !Kung San of Botswana (Blurton Jones, Draper, and Konner); the Hadza of Tanzania (Hawkes, Blurton-Jones, and O'Connell); and the Gabbra of

---

[1] The range of topics covered by human behavioural ecologists has been reviewed by Borgerhoff Mulder (1991), Cronk (1991), Voland (1998), and Winterhalder and Smith (2000).

Kenya (Mace). These populations live in a variety of habitats, and provide an opportunity to compare and contrast how human beings behave across a range of ecologies.

Key concepts

Human behavioural ecology is characterized by an emphasis on the flexibility of individual behaviour, by the testing of predictions derived from theoretical models, and by an initial assumption that human behaviour may consist of adaptive tradeoffs or compromises. These ideas are described in the following sections.

Flexibility of individual behaviour

A main assumption of human behavioural ecology is that human beings have been selected to optimize their lifetime reproductive success in response to environmental conditions by flexibly altering their behaviour. While people may exhibit a certain number of universal behaviour patterns across all environments, these researchers believe that most aspects of human behaviour are likely to depend in a facultative manner upon the particular social and ecological resources to which they are exposed. They suggest that a past history of selection will have favoured the ability to adopt the particular strategy that maximizes the difference between the benefits and costs in that particular environment. Such strategies may take the form 'in context X, do a; in context Y, switch to b' (Weinrich, 1977; Smith, 2000). Thus for human behavioural ecologists, our species is characterized by an extraordinary 'adaptability', a term used by evolutionary biologists to describe the degree to which a species can survive and successfully reproduce in a wide range of environments (Endler, 1986).

The idea of human beings as fitness-maximizing agents does not require that humans make a conscious decision to alter their behaviour in accordance with optimality criteria. Few people consciously calculate how to leave as many descendants as possible. Rather, human behavioural ecologists suggest that individuals may be predisposed to optimize their food acquisition rate or their social status possibly through conscious decision-making, and that these variables will correlate with an increased lifetime reproductive success. A key assumption is that a history of selection has endowed our species with a tendency to respond to the environments in which we find ourselves by weighing up the costs and benefits of adopting particular strategies. How specific cues from the environment result in a change in behaviour may depend upon physiological, psychological, and cultural influences. However, human behavioural ecologists believe that an understanding of the relative importance of these factors is not a requisite to studying the fitness outcomes of particular strategies.

Formal models and hypothesis testing

An important aspect of human behavioural ecology is the testing of hypotheses derived from formal or mathematical evolutionary theory. Starting with the assumption that behaviour has been selected to optimize reproductive success, models can be produced that predict the optimal pattern of behaviour in a given circumstance. The data collected on how human beings really behave are then compared to the model. Where the data fit the model, the hypothesis is upheld, which implies that the model provides a reasonably accurate description of the behavioural strategies that these people employ, the manner in which they make their decisions, and the particular cues to which they attend. Predictions can then be made about whether we might expect human beings in other environments to behave in a similar or different manner and data from different populations can be compared. Where the data do not fit, the model can be revised to include other variables or tradeoffs and re-tested, if necessary repeatedly, in a manner that allows researchers to home in on an understanding of the particular population. If the models remain a poor description, the conclusion may have to be drawn that there is no evidence to support the idea that human beings are behaving optimally in that situation.

An example of a body of theory that has been developed by behavioural ecologists is optimal foraging theory (Stephens and Krebs, 1986). This approach constitutes an attempt to construct models that specify a general set of decision rules for predators as they search for food. A central assumption is that foragers will have been naturally selected to make those choices that yield the greatest payoff; that is, the biggest difference between the benefits and costs of a particular strategy. While the ultimate function of a behavioural strategy is to optimize reproductive success, how particular strategies relate to changes in the number of viable offspring can be difficult to quantify. Instead, proximate currencies, such as the number of calories gained per hour of foraging, are used to estimate optimality, on the assumption that those individuals that forage most efficiently will on average leave the most descendants. Which currency is being optimized is not always clear and so different models can be developed to predict behaviour if individuals are maximizing the rate of energy intake, compared with, say, minimizing time spent foraging or the risk of predation. In each case, how a currency is optimized may depend upon factors such as the type of food item chosen by the individual, the time spent foraging on a particular resource, the choice of foraging patch, and the decision to forage alone or with other individuals. Assumptions about the constraints that limit the animal's feasible choices, such as its sensory abilities and the distributions of resources in the environment, are also built into the analysis. The costs and benefits of choosing particular strategies are then computed and the optimal strategy derived. The predictive performance of different models can then be compared to assess their relative merits.

Optimal foraging theory has been successfully utilized to investigate the feeding decisions of a broad range of animals (Stephens and Krebs, 1986), so it is no surprise that this body of theory should be applied to interpret human behaviour. We will discuss one such application further on in this chapter.

Adaptive tradeoffs

In the real world, the effective harvesting of energy from food is not the only problem that animals have to solve, and the competing demands on their time and resources are frequently in conflict (Stephens and Krebs, 1986). The best feeding site may also be teeming with dangerous predators, or a location may be good place for finding food but have nowhere suitable for a home. Also, a maximally effective diet may require animals to make a balanced choice of a range of prey items and not just grab as many calories as possible. Optimization models analyse how animals solve such tradeoffs. Because any 'unit of effort' can be invested only once, evolution is thought to have shaped animals into strategists that optimize how this investment is allocated into different aspects of their life history (Stearns, 1992). Behavioural ecologists are interested in how environmental factors influence the costs and benefits of particular tradeoffs (Voland, 1998). One example of a tradeoff is between somatic effort (growth of body tissue) and reproductive effort; i.e. to invest in one's self or to reproduce? A second is between direct and indirect reproduction; i.e. to reproduce oneself or to help relatives to reproduce? A third is how to balance mating and parental effort; i.e. to search for more mates or to invest in current offspring? A fourth tradeoff is between investment in quantity or quality of offspring.

An example of an adaptive tradeoff in animals is clutch size in birds, where parents have to strike a balance between producing too many and too few offspring. David Lack's classic books, *The Natural Regulation of Animal Numbers* (1954) and *Population Studies of Birds* (1966), provided detailed studies that investigated whether individual birds produce clutch sizes that are optimal. Beyond a certain point, larger clutch size will result in lowered overall parental reproductive success, as the number of chicks born increases but their survival rates decrease as the parents can no longer cope with feeding them all. In contrast, laying too few eggs represents a wasted opportunity if a pair could have raised more young. Lack found that intermediate clutch size yielded higher lifetime reproductive success than the physiological maximum because of limits on parents' ability to provide resources and care to offspring. Lack's work was highly influential in the establishment of behavioural ecology as a framework for studying animal behaviour. While humans do not generally have clutches of offspring, in the next section we will see how Lack's analysis greatly influenced thinking about how human parents may optimally adjust their family size or the inter-birth interval among their children.

The example of clutch size also illustrates an important limitation on using correlational data to determine tradeoffs. Simple models of optimality suggest that all individuals should aim for the same optimum and hence variation in the behaviour of individuals implies that some are behaving suboptimally. However, Lack (1954) pointed out that an individual's optimal strategy is likely to depend upon the resources available to that individual at the time. As some parents may be less likely to be able to invest in their offspring because they are still young or live on poor quality territories, the optimal clutch size for these parents may be lower than that for older individuals or those on better quality territories. Therefore, although one may predict a tradeoff between clutch size and number of surviving offspring for the population as a whole, such a negative correlation may not be found because the resources available will differ between individuals.

The problem of correlations between variables can be illustrated by a helpful example from the behaviour of human beings (van Noordwijk and de Jong, 1986). When money is a limiting resource, it can be spent on a car or a house, but not both, so for any given family there is a tradeoff. However, a survey of the values of cars and houses across households will generally reveal a positive (rather than the expected negative) correlation between amount spent on cars and houses. This is because households differ in the amount of money that they have and those with more money generally choose to increase the amount spent on both of the two commodities. In birds, it is possible to get round this complication through experimentation. For instance, the actual importance of territory quality can be investigated in animals experimentally by altering the clutch size of individuals on different territories to test whether the tradeoff between clutch size and offspring survival is influenced by territory quality. However, as such experimental manipulation is not possible with human populations, care should be taken in interpreting cor relations, or lack of correlations, between variables within a population, a problem often referred to as *phenotypic correlation*. Stronger tests of hypotheses can sometimes be made by looking at comparisons between populations rather than trying to explain variation within populations of individuals, or by controlling for one of the confounding variables within a population (Borgerhoff Mulder, 2000). State-dependent models also allow more sophisticated predictions to be made about how the optimal strategy may depend upon the state, or condition, of the individual (Mangel and Clark, 1988).

Case studies

This section provides examples of research in human behavioural ecology. We first describe the use of optimal foraging theory to study the hunting behaviour of human beings, then explore the relationship between marriage practices and environmental conditions. Finally, we will show how the tradeoff between offspring number and offspring quality has been studied among human beings and investigate the puzzle of the 'demographic transition'.

Foraging strategies and optimal group size

The Inuit population of Arctic Canada exploit a wide variety of animal species as food, ranging from fish and waterfowl to caribou and marine mammals, such as seals and beluga whales. They also use a number of different hunting methods, including traditional forms such as seal hunting at breathing holes. Foraging is often a solitary activity but hunting in groups is also a common occupation. Eric Alden Smith of the University of Washington carried out an investigation to see whether the Inuit foraged for food in an optimal manner. Smith (1985) was interested in questions such as 'why do they typically form groups of five to sixteen to hunt belugas and groups of two to ten to hunt seals, but hunt ptarmigan solitarily?' An obvious answer would be prey size; however, lake trout are often hunted by ten individuals while seals may occasionally be hunted by single hunters. Smith wondered whether foraging in groups may: (1) result in greater foraging success than would hunting alone; (2) result in a neutral or negative effect on an individual's hunting efficiency but provide other benefits, such as defence against predators or exchange of information; or (3) result simply from an aggregation of prey in a single location (Smith, 1985). One prediction from optimal foraging theory is that individuals will form groups of the size that maximizes the average capture rate of prey per hunter. To test this prediction, Smith collected data on hunting group sizes and amount of prey caught, and evaluated whether the most common group size for a particular prey type or hunting method was equal to the estimated optimal group size.

This case study indicates that when the data do not fit the assumptions of a simple model, the lack of fit can often be very informative in pointing out what might actually be going on. Smith's results were mixed. His data indicated that for prey that were most effectively caught by a single hunter, such a geese and ptarmigan, the most common foraging group size was indeed one. However, where more than one individual was generally required to capture a prey item, the most common group size was not found to be optimal. For example, breathing hole hunts provided the best returns per hunter when the hunting group size was three, but the most common size was four and could range up to eight individuals. Smith concluded that either the net capture rate was a poor currency to use to estimate optimal group size or, more interestingly perhaps, individuals were unable to maximize their net intake when foraging involved social interactions. Imagine a hunter that wanted to try to catch a seal at a breathing hole. If the hunter was very unlikely to catch a seal by hunting alone, he would attempt to join a hunting group to increase his chances of success. If a group of three hunters were about to leave on a foraging trip, the lone individual might try to go along with them. Taking the fourth hunter along would decrease the average amount of food that each hunter receives, and the members of the group would be predicted to attempt to leave without him. Any other lone hunters around would also probably want to combine with the group as long as they do better by joining the group than by hunting alone. If members of the group were able to stop other hunters from coming along, the average group size may remain at its optimum of three. However, if members were unable to refuse entry to others, the size of the group would continue to grow until the payoffs of hunting in a group were about the same as for hunting alone. The most common situation would usually be between these extremes, and depend upon the relative costs and benefits to joiners and members during such conflicts of interest. Smith's finding that the average group size for hunting seals at breathing holes was greater than his predicted optimum, suggested that optimization was constrained by other factors such as social interactions and the strategies of others. Strikingly similar results have been reported for hunting group size in lion populations, with group size being larger than predicted from an optimality model since the benefits to an individual of joining a group are greater than the benefits of hunting alone (Mangel and Clark, 1988).

Marriage practices

Human behavioural ecologists have long been interested in whether human mating patterns vary as a function of the local ecological context. A rare form of human marriage that has received considerable attention from anthropologists is that of polyandry, in which one woman is legally married to two or more men. Polyandry has been particularly well studied in the Zanskar and Ladakh regions of Tibet, where a number of brothers may be married to one wife. In the Himalayan villages of these regions, montane desert surrounds small areas of land that can only be cultivated because of the streams of snow-melt that flow down from the glaciers. Each family produces crops and raises animals on an estate that the eldest son usually inherits from his parents. The estate could be divided between the sons of the family; however, below a certain size, the divided farm would be too small to maintain a family. Instead, the brothers might benefit by jointly marrying a single wife and working the farm together.

John and Stamati Crook of Bristol and Oxford University, respectively, investigated the lives of these villagers and suggested that polyandry is functionally adaptive in the particularly harsh environment of these areas of Tibet (Crook and Crook, 1988). They discovered that women in polyandrous marriages had more offspring than women in monogamous marriages. However, did both brothers benefit from sharing a wife? Crook and Crook calculated that brothers would do well to remain in the polyandrous relationship, provided that they all had equal chance of fathering the offspring. However, this was unlikely to be the case, as eldest brothers usually had priority of access to their wife. Consequently, Crook and Crook predicted that younger brothers should seek monogamous relationships whenever possible and accept polyandry only when life circumstances constrained them to do so. They found that when alternative sources of income became available, many younger brothers left to search for a wife of their own. Smith's (1998) review of polyandrous marriage in Tibet concluded that younger brothers often do benefit from joining a polyandrous marriage with an older brother even where the paternity of offspring is skewed, given the alternative options. The observation that in the same regions of Tibet, town-dwellers generally exhibited monogamy and divided any wealth between their children suggested that the divisibility of the family wealth may be an important factor in determining what form of marriage is adopted. This case study indicates how human marriage practices may respond flexibly to the particular environmental circumstances in which particular individuals find themselves.

It is a general sociobiological adage that in many species a male's reproductive success can be greatly enhanced by mating with several females (Trivers, 1972). Among the many polygynous human societies, where resources and wealth can be monopolized, males are predicted to marry more than one wife if possible. In polygynous societies, wealth is more often transmitted to sons than to daughters, perhaps because inheritance of wealth by males is more likely to influence the number of grandchildren produced than is inheritance by daughters (Hartung, 1982). Payments for brides by prospective husbands or their families are also more commonly found in societies with inherited wealth and status, suggesting that males compete to use their resources to attract wives into polygynous marriages (Hartung, 1976; 1982). However, how might a female benefit from marrying an already married man? Beha-

vioural ecologists have suggested that a female may choose a
suitor who already has a partner if he has more than double
the resources of a bachelor (Orians, 1969). Female reproductive success has commonly been viewed as being dependent upon the amount of resources, rather than mates,
available to that female. The idea that, over a certain threshold, the amount of resources monopolized by a male will
make him more attractive than a competitor with no other
partners is referred to as the 'polygyny threshold model'
(Verner, 1964; Verner and Willson, 1966; Orians, 1969).

Monique Borgerhoff Mulder of the University of
California, Davis, investigated whether, given the choice
between marrying monogamously or polygynously,
Kipsigis women in Kenya adopted the alternative that
would lead to the greatest reproductive success (Borgerhoff
Mulder, 1990). In this population, men can marry up to
twelve wives though two, three, or four wives is a more
common number, and wives and children are dependent
upon the crops and animals produced on the husband's
land. These wives are obtained from their parents with a
'bridewealth' payment, the value of which is larger for
potential wives that appear fertile and have earlier onset of
menarche (menstrual cycling). The data indicated that,
given a choice of potential husbands, women chose to
marry the man able to provide the most resources to her,
upholding a prediction of the polygyny threshold model.
The partner chosen was not always the male with the largest
farm as the choice also depended on the number of wives
that the man had already. However, the number of wives
strongly correlated with a man's resource ownership. The
data also indicated that polygynously married Kipsigis
women had fewer surviving offspring than monogamously
married women. This result suggests that the women may

have been constrained in optimizing their choices, with one factor being that their marriage decisions are made with no information on how many other wives the husband is likely to take in the future.

While in some situations co-wives may cooperate to help each other, competition for greater investment in their own offspring may make relationships between co-wives antagonistic. More recent theories on mating systems in animals have highlighted the importance of conflict between individuals, including between male and female partners (Davies, 1989; Smuts and Smuts, 1993; Clutton-Brock and Parker, 1995; Westneat and Sargent, 1996). Reflecting these developments, human marriage patterns are also increasingly being studied as outcomes of conflicts of interest between individuals (e.g. Sellen *et al.*, 2000; Strassmann, 2000).

Tradeoffs between number and quality of offspring

Earlier we saw how clutch size in birds represents a tradeoff between the number of eggs laid and the quality of the young produced, measured in terms of fledgling survival. Human beings do not usually produce broods of offspring but there can be variation in other aspects of parental investment such as the length of interval between single births, usually known as the inter-birth interval. Too short an inter-birth interval can jeopardize the life of the younger infant. In an early study using the human behavioural ecology approach, Nick Blurton Jones of the University of California, Los Angeles, and colleagues investigated whether the inter-birth interval exhibited by a hunter–gatherer community, the !Kung San of southern Africa, optimized the number of surviving offspring produced by mothers (Blurton Jones, 1986). The research used models that included the estimated costs of raising children in this environment.

The !Kung San were intensively studied by anthropologist Richard Lee and ethnographer Nancy Howell in the 1970s. The detailed demographic data collected by Howell (1979) and Lee (1979) were used by Blurton Jones and colleagues to try to explain the long inter-birth interval exhibited in these populations. The !Kung San exhibit an average inter-birth interval of 4 years, unusually long for a population without modern contraceptives. Lee (1979) had suggested that the work entailed by shorter inter-birth intervals was simply too much for mothers as they would then be required to carry an infant, an older offspring, and all of the food collected during a foraging excursion. He also argued that a long inter-birth interval was an adaptation to the benefit of the mother by reducing her work effort, and to the population by limiting excessive population growth. Lee's explanation therefore combined ideas of group selection with those of individual selection.

Blurton Jones and Sibly (1978) attempted to model exactly how the workload of the mother would influence the optimal inter-birth interval. They predicted that short inter-birth intervals would be accompanied by more offspring deaths and, using the data collected by Lee, indeed found that an inter-birth interval much shorter than four years resulted in increased infant mortality (Blurton Jones, 1986). Mothers carried their children with them while foraging and allowed them to suckle until the age of around 4 years. Frequent suckling has been found to suppress ovulation in the mother and may have been partly responsible for the long interval to the following conception. This study was the first to apply evolutionary logic and an optimality approach to the study of human fertility.

Blurton Jones therefore found a positive correlation between offspring survival and inter-birth interval. The opposite prediction would also have been plausible: off-spring born after a short inter-birth interval may have been *more* likely to survive than those born after a long interval because their mothers were in good physical condition, a case of phenotypic correlation. Blurton Jones acknowledged that the prediction of a positive correlation between inter-birth interval and infant survival may have been overly simplistic but did fit with the data (Blurton Jones, 1997). Criticisms of Blurton Jones's study arose (e.g. Pennington and Harpending, 1988; Harpending, 1994) and rebuttals were published (Blurton Jones, 1994; 1997). However, attempts to replicate Blurton Jones's results with data on the Ache of Paraguay have failed to show that shorter inter-birth intervals were associated with higher infant mortality (Hill and Hurtado, 1996). Hill and Hurtado (1996) suggested that the body weight of the mother was a much more influential factor on infant survival.

In addition to the physical condition of the mother or length of the inter-birth interval, the wealth of the family may influence the number of offspring born. In a variety of pre-industrial societies, the wealth of the parents has been found to correlate positively with the number of offspring produced (reviewed by Kaplan and Lancaster, 2000; Low, 2000; although see Hill and Hurtado, 1996). Where the amount of wealth inherited by children has an important effect on their chances of marrying and producing grand-children, parents may limit the number of offspring that they have below the maximal rate. The tradeoff between number of offspring and the investment required by each offspring is increasingly being seen as an important limit on family size. Ruth Mace, an anthropologist at University

College London, has studied reproductive decisions of camel-herding Gabbra population of Kenya, where wealth is measurable in terms of numbers of camels and goats owned by the family (Mace, 1996). When deciding whether to have another baby, Gabbra parents appear to take into account the probability that they would be able to raise the child and marry it off successfully at maturity. In order to obtain a wife for a son, parents must give up part of their herd as well as making a bridewealth payment. Having to marry off too many sons could be of detriment to the rest of the family. Mace (1996) used a mathematical model to analyse the tradeoff between family size and marriage prospects, and showed that the decision as to whether to have another offspring depended upon the wealth of the family and the number of sons produced already.

The demographic transition

Although wealth of the parents has been found to correlate with number of offspring raised in many human populations, this correlation does not hold for post-industrial societies, in which wealthier families often have fewer offspring than poorer families. Countries in which wealth and family size do not correlate positively have generally passed through a 'demographic transition'; that is, a period of history in which dramatic changes in fertility and mortality have occurred along with a rise in living standards (Borgerhoff Mulder, 1998). An example occurred in 19th-century Europe following the Industrial Revolution. Such transitions are generally characterized by a decline in mortality and a radical decline in the number of children that parents produce. Another feature of demographic transitions is that rich families reduce their fertility earlier, and often more markedly, than the rest of the population. Why would people choose to limit their reproduction voluntarily when resources are apparently plentiful? This lack of a positive correlation between wealth and number of offspring is held up by some as proof that humans no longer behave in a manner that optimizes their reproductive success and that evolutionary approaches do not explain current family sizes (Vining, 1986).

Traditional methods of historical and economic demography, however, have failed to develop a robust explanation for demographic transitions. Human behavioural ecologists have provided an alternative viewpoint for investigating why this pattern of fertility change may occur (reviewed by Borgerhoff Mulder, 1998). Early hypotheses suggested that having fewer offspring may be an optimal strategy in modern societies where high levels of investment in offspring are critical to the child's success and costly to the parents (Irons, 1983; Turke, 1989). By providing more resources to fewer offspring, wealthy parents may increase the number of grandchildren that are produced, if the money spent on education, for instance, greatly enhances the child's ability to reproduce successfully in the future (Kaplan, 1996; Kaplan and Lancaster, 2000). From this viewpoint, the negative correlation between wealth and fertility may be seen as an adaptive strategy. Mace suggested that parents in different social strata use different decision rules regarding family size, so that within each social stratum wealth still correlates with number of offspring produced (Mace, 2000). However, there is little evidence available to suggest that wealthy parents always optimize their lifetime reproductive success by having few offspring (Rogers, 1990; Kaplan et al., 1995). Further modelling of tradeoffs may provide information on whether parents are using more complex investment strategies, such as optimizing the amount of heritable wealth per child (Luttbeg *et al.*, 2000).

Perhaps human beings may not have evolved the capacity to optimize tradeoffs that involve disparate currencies such as money, land, and cattle, and therefore are not able to assess accurately the choices available to them and make optimal decisions about the number of children they can have (Kaplan and Lancaster, 2000). Alternatively, the reduced levels of fertility seen in modern societies may be a response to the absence of close kin networks. The presence of kin may greatly influence the ability of mothers to raise children in traditional societies (Turke, 1989; Draper, 1989). Draper (1989) argued that human beings place greater importance on the presence of kin than the availability of monetary resources as a cue for how many children to raise to that, when parents perceive an absence of helpers, they produce fewer offspring than they actually could. Draper's account of fertility decisions in modern societies is a good example of an explanation that integrates ideas derived from both human behavioural ecology and evolutionary psychology (see Chapter 5).

In comparison to these studies, Hill and Hurtado (1996) have suggested that it is premature to assume that low fertility in modern societies is due to the presence of new factors that were never historically experienced by human populations. Fertility levels that are lower than those that would apparently maximize long-term fitness appear to characterize some traditional societies and not just modern populations (Hill and Hurtado, 1996). Fertility levels in all populations may be sensitive to parameters such as infant and juvenile mortality risks and the unpredictability of food. Hill and Hurtado (1996) suggest that considerable theoretical developments will be required to provide more general models of human fertility patterns. This area of research may benefit from the inclusion of cultural process-es into the models, and hence the incorporation of factors such as the presence of contraceptives and the increased opportunities for women to be financially independent.

Critical evaluation

In the late 1980s, human behavioural ecology came under attack from the founders of the newly forming discipline of evolutionary psychology. The most hostile remarks came from Donald Symons (1987) at the University of California at Santa Barbara, in an infamous essay entitled 'If we're all Darwinians, what's the fuss about?' Symons argued force-fully that the research programme of the human behav-ioural ecologists was seriously misguided because it did not formulate or test hypotheses concerning human adaptations or shed light on the human mind where such adaptations would be found, but merely established which behaviour patterns appeared adaptive by correlating human behav-ioural traits with reproductive success (Symons, 1987; 1989). Confusingly, two very similar terms have been used for two distinct ideas. An *adaptation* is a character favoured by natural selection for its effectiveness in a particular role; that is, it has an evolutionary history of selection. To be labelled as *adaptive*, a character has to function currently to increase reproductive success. As Figure 4.1 illustrates, not only are adaptive traits not the same as adaptations, but they can be regarded as being independent.

For Symons (1990, p. 430) 'correlating trait variation with reproductive success' was 'an ineffective, ambiguous, and inconclusive way to study adaptation'. He stressed how human characteristics that currently appear adaptive might

133

	Is the behaviour adaptive? *Adaptive behaviour* is functional behaviour that increments reproductive success	
	Yes	No
Is the behaviour an adaptation? An *adaptation* is a character favoured by natural selection for its effectiveness in a particular role — Yes	**Current adaptation** A current adaptation is an adaptation that has remained adaptive because of continuity in the selective environment	**Past adaptation** A past adaptation is an adaptation that is no longer adaptive because of a change in the selective environment
No	**Exaptation** An exaptation is a character that now enhances fitness but was not built by natural selection for its current role	**Dysfunctional by product** A dysfunctional by-product is a character that neither enhances fitness nor was built by natural selection

Figure 4.1 The difference between adaptive behaviour and adaptations.

nonetheless not be adaptations. One reason for this discrepancy was pointed out by George Williams in his classic 1966 book *Adaptation and Natural Selection*. Williams emphasized how it is important to distinguish adaptations from characters with fortuitous effects. The latter, which Gould and Vrba (1982) subsequently termed 'exaptations', are features that now enhance fitness but were not built by natural selection for their current role. For instance, Williams wrote 'I cannot readily accept the idea that [human] advance mental capabilities have ever been directly favoured by selection' although they 'might possibly be produced as an incidental effect of selection for the ability to understand and remember simple verbal instructions early in life' (1966, pp. 14–15). Here, language would be an adaptation while human intelligence is an exaptation.

Symons maintained that many human adaptations might not be currently adaptive, but rather were adaptations to a bygone world inhabited by our ancestors. For instance, whether or not it is currently adaptive, the human taste for sugar and fat in the diet is an adaptation to a past hunter–gatherer existence where one couldn't get enough of these energy-rich nutrients, so that they are pleasurable to us. Importantly, Symons also argued that the adaptations that underlay human behaviour were to be found at the psychological level: they were the cognitive machinery that control behaviour, which the human behavioural ecologists largely ignore. For Symons, the best way to use evolution to study human behaviour is to take an *adaptationist* approach; that is, to search for the psychological mechanisms that constituted the adaptations that regulate behaviour, rather than an *adaptivist* approach that identifies adaptive behaviour among humans but which may bear no relationship to human adaptations. He wrote:

Darwinism is a historical explanation of the origin and
maintenance of *adaptations*, and almost none of the
phenomena of interest to social scientists—polyandry,
bridewealth, the avunculate, and so forth—are themselves
adaptations. Whether or not they are adaptive, they cannot
be adaptations because they are not descriptions of
phenotypic design. Darwinism can be 'applied' to
traditional social science phenomena only insofar as it
illuminates the psychological adaptations that underpin
those phenomena. (Symons, 1990, p. 435)

Other evolutionary psychologists, notably Cosmides and
Tooby, joined in Symons's attack (Cosmides and Tooby,
1987; Tooby and Cosmides, 1990a), and it was partly this
collective onslaught that launched evolutionary psychology
as a distinct field in its own right, as we discuss in the following chapter. Not surprisingly, the human behavioural
ecologists defended their position, and a vigorous and
sometimes bitter debate ensued that continues to this day.

In the following sections we evaluate the criticisms of
human behavioural ecology. There are various components
to the salvo from evolutionary psychology, which we will
consider in turn. They include whether it is better to focus
on human behaviour or on psychological mechanisms, the
relative merits of adaptationism and adaptivism, and the
possibility of suboptimal behaviour. We will then go on to
consider a fourth line of criticism, which emanates more
from the social science community, which questions the
legitimacy of studying human behaviour and institutions in
a piecemeal fashion.

Behaviour *versus* psychological mechanisms

Are the evolutionary psychologists correct in their claim
that analysis of psychological mechanisms is the most

appropriate level at which to look for human adaptations? Or are the human behavioural ecologists justified in focusing on adaptive human behaviour?

Clearly there are strengths and weaknesses to both perspectives. The human behavioural ecology position has the advantage that behaviour can be easily observed and recorded, it can be studied in a rigorous scientific manner, and it can be modelled with formal mathematical theory. Such analyses may not tell researchers much about the mind or human psychological adaptations, but since this is not the goal of such studies it is of little relevance. To the extent that the theoretical stance adopted by human behavioural ecologists leads them to a deeper understanding of human behaviour and institutions, their approach is vindicated. Even the small number of studies reported here are sufficient to draw the conclusion that human behavioural ecology has provided valuable and important insights.

Far from lamenting the fact that their approach pays little attention to proximate mechanisms, many human behavioural ecologists regard this as a virtue. For instance, Smith (2000) is open about the human behavioural ecologist's adherence to the *phenotypic gambit* (Grafen, 1984), which posits that the constraints on human adaptiveness, be they genetic, psychological, or social, are so minimal as to justify their being ignored in the construction of models and the testing of hypotheses. For many human behavioural ecologists, it simply doesn't matter whether humans end up behaving in an adaptive manner as a consequence of their psychological mechanisms, their learning or their culture. So long as their behaviour is adaptive, then it can be predicted with formal models. For these researchers, the key legacy of our evolutionary history is *adaptability* not psychological or behavioural adaptations. This adaptability must itself be an adaptation, albeit an extremely general one (Borgerhoff Mulder *et al.*, 1997).

Moreover, notwithstanding the open adherence to the phenotypic gambit, there are grounds to dispute the claim that the behavioural ecology approach ignores proximate mechanisms. Behavioural ecologists frequently engage in the recursive exercise of model building, testing the model against human or animal data, and then revising the model if there are discrepancies in its performance, until such a time as there is a good fit between model and data. When successful, this exercise allows researchers to home in on the behavioural strategies that their subject organisms are employing. For instance, they may learn that their animals are maximizing the net rate of caloric intake by attending to various cues about prey items and their location, and utilizing specific rules of thumb to guide them as to the optimal choice of prey and to the best way to move between foraging patches. The formal models represent hypotheses about how cues are weighted in different contexts, how they are evaluated statistically in the brain, and how they are used in the real world of decision-making (for an example see Luttbeg *et al.*, 2000). In other words, while behavioural ecologists do not start with a focus on proximate mechanisms or psychological constructs, the analyses in which they engage can rarely be carried out successfully without redress to such mechanisms and constructs, and frequently have an understanding of such mechanisms as their goal.

Symons (1987; 1989) and Cosmides and Tooby (1987) argue that natural selection cannot act directly on behaviour, but rather on the behavioural regulatory machinery that underpins it. Hence, they maintain that the adaptations that are expressed in human behaviour will mainly be found at the psychological level. Human behavioural ecologists have responded in a variety of ways to this charge. Turke (1990) asserts that natural selection acts on all aspects of the phenotype from the physiological to the behavioural, and thus there is no reason to focus on the psychological level. Alexander (1979) and Borgerhoff Mulder (1991) claim that natural selection will choose between alternative sets of 'decision rules' and select those sets of rules which give the best outcome in terms of generating those behaviour patterns which maximize lifetime reproductive success. In contrast, psychological mechanisms are poorly defined and understood, and are difficult to identify in field conditions. A third refutation is to respond to the criticism as if it were water off a duck's back, an indifference stemming from an ambivalence to the causal processes that underlie adaptive behaviour. Who is right on this contentious issue is a matter of personal judgement. The assertion from evolutionary psychology that selection acts on psychological mechanisms is based on deductive reasoning rather than empirical data, but then so are the counterclaims of the human behavioural ecologists.

We shall see in the next chapter that the evolutionary psychologists have accumulated evidence that there are human psychological adaptations. To the extent that bona fide psychological adaptations of the kind that evolutionary psychologists propose exist (i.e. highly specified and domain-specific structures), then the human behavioural ecologists' phenotypic gambit may be compromised. This is because evolved psychological mechanisms may constitute constraints on human flexibility and our behaviour may be channelled in certain directions by them. Yet there is no reason to regard this as a fundamental challenge to human behavioural ecology. On the contrary, some human behavioural ecologists accept that the inclusion of cognitive mechanisms that produce behavioural output would be of interest (e.g. Smith, 1992; Borgerhoff Mulder, 1998; Kaplan and Lancaster, 2000). Clearly there are limits on the flexibility of human behavioural response and these limits are of importance in investigating why human beings engage in one range of strategies rather than another. Hopefully, one positive outcome of the sometimes vitriolic debate between the evolutionary psychologists and human behavioural ecologists will be a fruitful integration of concepts.

The adaptationist–adaptivist debate

Evolutionary psychologists (e.g. Symons, 1987; Tooby and Cosmides, 1990a) charge human behavioural ecologists with confusing adaptive behaviour with adaptations, and with neglecting the distinction between adaptations and exaptations (see Figure 4.1). To what extent this charge is justified is difficult for us to judge. Certainly, the human behavioural ecologists to whom we have spoken are well aware of these distinctions and acknowledge that correlations between behaviour and current reproductive success can only *suggest* hypotheses about adaptation (Caro and Borgerhoff Mulder, 1987). Yet some of the published literature appears implicitly to assume that to demonstrate that a trait is adaptive is sufficient to be able to conclude that it has an evolutionary explanation.

An example that illustrates the distinction between the study of current function and the evolutionary history of a trait is the recent discussion of the evolution of menopause. The puzzle of menopause is encapsulated in the question 'wouldn't women increase their reproductive success by continuing to produce offspring until they die?' An early idea, known as the 'grandmothering hypothesis' (Williams, 1957), suggested that greater benefits may be gained by women through investment in already existing children and in grandchildren than could be expected from continuing to invest in the production of additional offspring. Some behavioural ecologists suggest that, if grandmothers could be shown to do better by investing in grandchildren than in producing more of their own offspring, menopause could be explained by a history of selection for grandmothering behaviour. Grandmothers have been shown to play an important role in feeding and caring for grandchildren, for example in the Hadza of Tanzania (Hawkes *et al.*, 1989). However, describing a current functional benefit does not provide an account of past evolutionary history (Hill and Hurtado, 1997). In fact, there are four possible ways to describe menopause which would fit into Figure 4.1. If menopause is an adaptation, the selection pressures that favoured it in the past may still be acting in the present (it is a current adaptation) or selection pressures may have changed so that menopause no longer has a selection advantage (it is a past adaptation). These two scenarios are shown in the top two boxes of Figure 4.1. Alternatively, menopause may not have been specifically selected in the past (therefore is not an adaptation) and it either has no beneficial effect in the present (a dysfunctional by-product) or provides some new benefit in the present environment that renders it functional and adaptive (in which case it is an exaptation). These two scenarios are shown in the bottom two boxes in Figure 4.1.

Human behavioural ecologists attempted to test the grandmother hypothesis by modelling the tradeoff between investment in grandchildren and production of further offspring. In fact, their models failed to show that the benefits of grandmothering behaviour would be sufficient to offset the potential benefits of having further

children (Hill and Hurtado, 1991; Rogers, 1993). Therefore, even theoretically, menopause could not be explained
as an adaptation using this hypothesis, However, alternative explanations for menopause as an adaptation are more
feasible. For example, termination of reproduction before
the end of life may be favoured if this helps the mother to
wean her final infant before she dies (Sherman, 1998).
Alternatively, menopause may not be an adaptation but is
perhaps an artefact of recent advances in medicine and
healthcare that have artificially lengthened the lifespan
(Washburn, 1981; Weiss, 1981), or a by-product of selection for fast reproduction early in life and an inevitable
part of the senescence process (Williams, 1957; Hill and
Hurtado, 1991; Packer *et al.*, 1998).

However, once menopause had arisen for whatever reasons, selection could favour other traits, such as grandmothering behaviour, that may have been unlikely to occur
in the absence of menopause. Human behavioural ecologists have now turned to the question of grandmothering
behaviour itself as an adaptation—'once a woman has
passed through menopause, will selection favour her
investing in grandchildren rather than providing no help?'
Also, as females of other primate species exhibit signs of
reproductive senescence after a certain age (Caro *et al.*,
1995), human behavioural ecologists have suggested that
menopause is not a derived trait within the hominids and
rather than it is the extended female lifespan after
menopause that requires explanation (Hawkes *et al.*, 1998;
2000). Grandmothering behaviour and the extended
postreproductive lifespan then become the traits under
investigation.

Evolutionary psychologists are correct in pointing out
that studying current function does not necessarily provide

any information about whether that trait was brought about by selection for that particular function. However, if current selection acts on heritable variation in that trait, the trait will have an evolutionary present. Technically, this would transform an exaptation into an adaptation (Endler, 1986a). Evolutionary psychologists are interested in the adaptations that influence human behaviour (and so concentrate on the top two boxes in Figure 4.1), but believe that many, perhaps even most, are no longer adaptive (past adaptations). The discrepancy between past and current environments may produce a mismatch between behaviour and the environment, known as an 'adaptive lag'. Their critics suggest that evolutionary psychologists under-estimate both the amount of currently adaptive human behaviour and, more specifically, the frequency of current adaptations. In contrast, the human behavioural ecologists begin with the assumption that behaviour patterns may have a current adaptive function (two left-hand boxes in Figure 4.1), although they need to remain aware of the distinction between current adaptations and exaptations, and of the existence of adaptations to past environments. There is no doubt that a complete evolutionary account would have to involve relating any observed adaptive or non-adaptive behaviour to the operation of an underlying adaptation and showing how the adaptation was adaptive in ancestral environments. Unfortunately, in the absence of a detailed knowledge of our evolutionary past, that is not always easy to do.

The fact that many human behavioural ecologists do not subscribe to the adaptationist programme, and appear content to assume that natural selection would have fashioned human behavior to be optimal within the constraints on the system led John Tooby to assert:

The study of adaptiveness merely draws metaphorical inspiration from Darwinism, whereas the study of adaptation *is* Darwinian.[2]

If the observation that a human behaviour is adaptive does not imply that it is a human adaptation, does that render the exercise of measuring the reproductive success of individuals worthless? Clearly not, since the primary and fundamental justification for their approach, given by human behavioural ecologists, is that it helps to explain variation in human behaviour. But does measuring reproductive success tell us anything about human evolution?

While correlating trait variation with reproductive success is insufficient to assert that a behaviour is an adaptation, it is nonetheless a fundamental tool in the evolutionary biologist's tool-kit (Endler, 1986a). It allows researchers to find out whether and how a species is evolving, and explore the characteristics of the evolutionary process. The assumption that the present is the key to the past is a feature of much contemporary evolutionary thinking. A knowledge of how selection is operating now allows us to extrapolate back in time and devise hypotheses concerning evolutionary trajectories on the assumption that selection pressures have remained constant, or have changed in a predictable manner. Of course, reconstructing such evolutionary account is problematic and researchers are vulnerable to concocting fanciful stories. However, such stories are even more likely to be fanciful if they are unconstrained by knowledge of current selection. What happens

---

[2] Quoted in Symons (1990, p. 427).

in the present informs our understanding of what happens in the past at least as much as vice versa, if for no other reason than the present is more visible and more subject to experimentation (Turke, 1990).

If there is any meaningful sense in which humans exhibit underlying psychological or behavioural adaptations, then there are two kinds of adaptationist hypotheses that need to be distinguished: those that predict adaptive outcomes and those that do not. Given sufficient environmental continuity between past and present selective environments, or sufficient flexibility in the behavioural regulatory procedures, adaptations are expected to produce adaptive outcomes. If, as evolutionary psychologists anticipate, the modern selective environment is very different from that in which human adaptations were forged, and psychological adaptations are highly specific, then adaptations may not produce adaptive outcomes. However, since no-one really knows to what extent the past and present selective environments differ from that trait, it is entirely possible that most human adaptations could produce adaptive behaviour in the modern environment, and it would be premature to assume that most would not. Humans are particularly adept at constructing their niche and hence it is even conceivable that the modern world has actually been fashioned by us to suit our psychological and behavioural adaptations (Laland et al., 2000), a hypothesis that would mean that the amount of 'adaptive lag' has been greatly overestimated.

Turke (1990) argues that knowledge of the contexts in which people behave adaptively may have the additional benefit of providing important information about the nature of the mechanisms that comprise the human psyche. He asserts that finding that an adaptation produces an adaptive outcome in particular environments but not in others may illuminate the selective background and ontogeny of that trait, including the extent to which the trait is specialized or has a general function. It remains to be seen whether Turke's claims can be verified, but for the moment this would seem a further plausible justification for the behavioural ecology approach.

The possible of suboptimal behaviour

Human behavioural ecology starts with the notion that human behavioural strategies have been shaped by selection to optimize reproductive success in particular environments: the actual data from human populations are then compared to predictions made from theoretical models. Where the data do not fit the model, there are two obvious explanations. First, the assumptions about the behavioural strategies being optimized or the estimates of the costs and benefits of particular strategies may be incorrect, or the model may not have incorporated the appropriate tradeoffs. Secondly, human beings may not be behaving optimally. However, if frequently appears to the outsider that human behavioural ecologists are reluctant to draw this second conclusion. Amongst the human behavioural ecology community there appears to be a certain kudos credited to the resourceful researchers who can show that human behaviour that hitherto appeared puzzlingly suboptimal and defied previous explanation is actually an optimal strategy. Given the recursive nature of the human behavioural ecology approach, it is understandable that these researchers should not want to admit defeat and conclude that behaviour is suboptimal prematurely, and might be tempted to try one further attempt at model fitting. As pointed out by Maynard Smith (1978), the role of optimization theories within biology is not to demonstrate that the organism is behaving adaptively but to use the assumption of adaptive behaviour as a tool to develop an understanding of the diversity of behavioural strategies. However, if researchers were to continue endlessly to move between model, data collection, and new model, and were never prepared to reach the conclusion that a particular behaviour might be maladaptive, this would mean, contra Maynard Smith, that the goal of the exercise was effectively to show that behaviour is adaptive.

The rarity with which human behavioural ecologists admit to a case of suboptimal behaviour has been further fuel to their critics. Symons (1990) gives the example of the Efe Pygmies of the Ituri Forest in the Democratic Republic of Congo, many of whom have recently taken up smoking at considerable personal cost in terms of time, energy, and money. He describes how among Efe men smoking and material wealth are negatively correlated, while material wealth and obtaining wives are positively correlated. Symons writes:

> To the adaptivist data like these are a theoretical challenge; the typical adaptivist's response to such a challenge is to cast about for some *ad hoc* reason why apparently maladaptive behaviour might conceivably be more adaptive than it seems. (1990, p. 433)

There are several theoretical grounds on which to suspect that human behaviour may sometimes be suboptimal. Evolutionary psychologists stress how modern situations are vastly different from past selective environments, frequently rendering our adaptations obsolete (Cosmides and Tooby, 1987). While human behavioural ecologists claim that humans exhibit relatively little adaptive lag, it is noticeable that their research rarely attends to the behaviour of Westernized populations. This is perhaps a tacit acknowledgement that human behaviour may not be adaptive in modern society and that optimality is more likely to be shown in populations exposed to more 'naturalistic' conditions. Formal analyses carried out by researchers in the gene–culture coevolution (or dual inheritance) tradition described in Chapter 7 have demonstrated that there are a number of means by which suboptimal behaviour can be favoured when genes and culture interact (Feldman and Laland, 1996). Evolutionary biologists commonly find in theoretical and empirical analyses that populations may get trapped at local optima which, where there are multiple fitness peaks, may prevent global optimization (Hartl and Clark, 1989). Natural selection is like a train that can only carry passengers uphill, and hence passengers may only reach the summit of the local hill rather than the highest peak in the region. Finally, humans are unlikely to be infinitely flexible and there may well be significant genetic and developmental constraints or predispositions that prevent humans from maximizing fitness under all circumstances.

Most human behavioural ecologists are willing to consider the possibility of suboptimal behaviour, at least theoretically (Smith, 2000). However, they also point out that much of human behaviour does fit with the predictions made by their models, and maintain that the assumption that human beings are selected to behave in ways that maximize their reproductive success is a useful starting point for studying human behaviour. Cases in which the data do not fit the predictions of the model can be as informative as cases in which a perfect fit is produced. For example, studies on the hunting behaviour of Ache men revealed that their behaviour did not fit that predicted by optimal foraging theory; they were spending more time hunting for meat items than predicted (Hill, 1988). Kim Hill (1988) suggested that, as well as calorific content, measures of nutrient content needed to be included into the model, while Kristen Hawkes (1991) later considered the possibility that men were hunting more often than predicted since benefits in terms of gaining matings was also a factor. The data on Ache men did indeed suggest that the best hunters gained more extramarital matings than poor hunters (Kaplan and Hill, 1985). Thus, even when the models fail, light is frequently thrown on the phenomenon in question.

Moreover, supposing that researchers suspect that a human behaviour pattern is genuinely suboptimal. What can they do with this conclusion? How could they test it definitively? Merely showing that it is suboptimal according to the criteria of one particular model always leaves open the possibility that alternative models will reveal this to be a false negative conclusion. What is needed is the technology to construct specific, testable hypotheses about suboptimal behaviour. Potentially, such methods could be developed through an integration of the formal models from behavioural ecology and gene–culture coevolution. In the absence of these technical advances, human behavioural ecologists would seem justified in carrying on in their current vein.

Piecemeal approach

The anthropological community is dominated by holistic approaches to understanding human behaviour and has been critical of the piecemeal method adopted by the human behavioural ecologists and other evolutionary minded researchers (e.g. Bloch, 2000). According to Smith (2000):

> The piecemeal approach holds that complex socioecological phenomena are fruitfully studied piece by piece—in a reductionist rather than holistic fashion. Thus, a complex

problem such as explaining the marriage patterns in a population is broken down into a set of component decisions and constraints such as the female preferences for mate characteristics, male preferences, the distribution of these characteristics in the population, the ecological and historical determinants of this distribution, and so on.

How legitimate is it to assume that humans are optimizing only one aspect of behaviour and that complex behaviour can be analysed piece by piece? Interrelationships between human institutions, such as kinship, law, and religion, means that many variables will need to be considered to gain a complete understanding and individual's behaviour (Hinde, 1987). Are there not likely to be trade-offs between a number of important currencies, such as foraging success, social status, and mate choice?

Our view is that the piecemeal approach, like all reductionism in science, is one necessary, pragmatic stance for dealing with complex phenomena. It is not possible to construct useful, analysable theoretical models without making simplifying assumptions. Such deliberate simplification is usually a virtue rather than a vice, to the extent that if focuses researchers' attention on the key processes that underlie the system, and removes less relevant but obfuscating factors. Simple models can always be extended to relax their assumptions, and analyses can always be broadened to incorporate additional variables. Mathematical modelling is a dynamic process. The most effective means to proceed is frequently to start with a simple model that concentrates solely on the most central processes and to elaborate gradually. The conclusion that the exercise is doomed by the inextricable complexity of human institutions is refuted by the many examples where the approach has worked (see Borgerhoff Mulder, 1991; Voland, 1998).

Conclusions

Most anthropologists and other social scientists are sceptical about, if not downright hostile to, the evolutionary
perspective of the human behavioural ecologists. Indeed,
the current post-modernist malaise that afflicts much of the
social sciences solicits a fashionably anti-science nega-
tivism. As we saw in Chapter 2, part of the hostility stems
from an acute awareness of the past abuses of Darwinism.
Unfortunately, some branches of the social sciences would
appear to have constructed their own straw-man version of
evolutionary theory that few biologists would recognize,
disparagingly labelled 'evolution*ism*', which tragically is
exemplified by the eugenics movement, sterilization laws,
racist immigration policy, and the evil biological rantings
of Hitler, rather than the countless positive ramifications of
Darwinism. We must remember too that anthropology as a
discipline was forged in an atmosphere dominated by the
erroneously linear and progressive 'evolutionary' doctrines
of 19th-century intellectuals such as Herbert Spencer,
Edward Tylor, John Lubbock, and Lewis Henry Morgan,
which fuelled racist ideologies. Once bitten, anthropologists remain shy of evolutionary reasoning. Thus, while
the methods of human behavioural ecology have the advantage that they are quantitative, rigorous, theory-driven,
and insightful, such qualities sadly are rarely appreciated by
the anthropological community at large, few of whom have
a mathematical training. As a consequence, despite the rich
vein of good ideas that have emerged from human behavioural ecology, and which are manifest in several hundred
scholarly publications, the approach remains a very small
branch of anthropology.

In the next chapter we will turn our attention to the rapidly growing field of evolutionary psychology, stripping it down to its bare-bone essentials, and critically evaluating its evolutionary credentials. In terms of the number of researchers, human behavioural ecology is dwarfed by its cousin evolutionary psychology. As we have seen, the two dominant evolutionary approaches to studying human behaviour have been involved in a sometimes-heated debate for many years now. Is this just a petty squabble between rival factions competing for territorial dominance or do the differences reflect genuine philosophical or methodological disagreements as to the best way to use evolution to interpret human behaviour? To answer these questions we need to take a closer look at evolutionary psychology.

CHAPTER 5

Evolutionary psychology

Researchers committed to an evolutionary perspective on humanity were initially united in the face of widespread hostility to human sociobiology. However, in the 1980s, as the number of investigators using evolution to study human behaviour increased, subgroups began to emerge with different opinions on how best to proceed. One such subgroup was dominated by academic psychologists searching for the *evolved psychological mechanisms* that they envisaged underpinned any universal mental and behavioural characteristics of humanity. While the intellectual roots of some of these practitioners could be traced to human sociobiology, or to the study of animal behaviour, the majority were fresh recruits who sought to differentiate themselves from human sociobiology, and restyled themselves as *Darwinian* or *evolutionary psychologists*. For Leda Cosmides and John Tooby, two of the pioneers of this new discipline, evolutionary psychology owed little intellectual debt to Edward Wilson but did draw inspiration from the writings of Bill Hamilton, Robert Trivers, and George Williams. Tooby, a Harvard-trained anthropologist who had worked closely with Irven DeVore, and Cosmides, a psychologist also from Harvard, were brought by Donald Symons to Santa Barbara where they founded the first Center for Research in Evolutionary Psychology.

The 'Santa Barbara school'[1] were concerned that human sociobiologists and behavioural ecologists had neglected psychological adaptations:

> In the rush to apply evolutionary insights to a science of human behavior, many researchers have made a conceptual 'wrong turn', leaving a gap in the evolutionary approach that has limited its effectiveness. This wrong turn has consisted of attempting to apply evolutionary theory directly to the level of manifest behavior, rather than using it as a heuristic guide for the discovery of innate psychological mechanisms (Cosmides and Tooby, 1987, pp. 278–9).

The evolutionary psychologists stressed how the environments that contemporary human populations experience differ massively from those experienced by our ancestors. Modern houses, cities, and social institutions are relatively recent innovations in evolutionary terms, and hence they suggested that there is a mismatch between our ancient psychological adaptations and our modern, artificially constructed world. As a result of this mismatch, they argued, researchers should not expect human behaviour to be adaptive. For evolutionary psychologists, any failure on the part of human sociobiologists and human behavioural ecologists to find optimal human behaviour would only demonstrate that these researchers were working at the wrong level (Symons, 1987).

Nevertheless, if evolutionary psychologists are correct in their reasoning that human beings walk around with stone-age minds in their heads, then the manner in which people

---

[1]   As Tooby, Cosmides, and Symons all work at the University of California at Santa Barbara, this strand of evolutionary psychology has become known (perhaps somewhat disparagingly) as The Santa Barbara Church.

think should betray their ancestral selective environments. They proposed that evolutionary biology was best used to generate hypotheses of the adaptive problems that the human mind had to solve in the selective environment of our ancestors. Following Bowlby (1969), this past environment was described as the *environment of evolutionary adaptedness* (EEA), which was generally conceived of as the Pleistocene[2] environment inhabited by our Stone-Age hunter–gatherer ancestors. With a good understanding of these adaptive problems, evolutionary-minded researchers would be able to determine the design features that any cognitive programme must have to be capable of solving them. This would help them to develop models of how the mind works. Thus, with evolutionary psychology, the primary focus of attention shifted from behavioural adaptations to evolved psychological mechanisms.

The evolutionary psychologists' approach was also influenced by the changing face of psychology which, by the 1980s, had long abandoned behaviourism and was in the throws of the cognitive revolution. The use of animals as research tools had been jettisoned in favour of the computer as an analogue of human cognition. Minds could be described in terms of information processing in which representations of the world were constructed on the basis of information from sensory inputs, while cognitive decision rules determined motor outputs. Research into artificial intelligence revealed that, to solve even supposedly simple cognitive tasks, minds required pre-specified procedures or information. This led evolutionary psycho-

---

2   The Pleistocene is the period from 1.7 million to 10,000 years ago.

logists to propose that 'innate psychological mechanisms' guided decision-making. Psychologists were increasingly developing computational theories of informational processing problems that specified what had to happen if a particular function was to be accomplished (Marr, 1982). Evolutionary psychologists believed that, with sufficient information about our ancestors' way of life, evolutionary theory could be put to use to construct computational theories of adaptive information processing problems.

Cosmides and Tooby's visionary writings were to provide the defining features of the field, and trigger the rapid growth of this new movement. By the 1990s evolutionary psychology had blossomed into a thriving programme of research, with important contributions from Jerome Barkow, David Buss, Bruce Ellis, Martin Daly and Margo Wilson, Steven Pinker, Roger Shepard, Donald Symons, and many others. With the publication of Barkow, Cosmides, and Tooby's (1992) landmark volume *The Adapted Mind,* a stream of popular books in this new genre followed, notably David Buss's (1994) *The Evolution of Desire,* Robert Wright's (1994) *The Moral Animal,* and Steven Pinker's (1997) *How the Mind Works.*

As noted in the introduction, however, the term 'evolutionary psychology' is used in a divergent manner by different researchers. Confusingly, some anthropologists or archaeologists describe themselves as doing 'evolutionary psychology' because they identify with the Santa Barbara perspective. Conversely, prominent evolutionarily minded psychologists, such as Henry Plotkin (1994, 1997), disagree with the modular and adaptationist school of thought championed at Santa Barbara. Many researchers have endeavoured to broaden evolutionary psychology to encompass all evolutionary approaches to the study of human minds and behaviour (Daly and Wilson, 1999; Buss, 1999; Barrett *et al.*, 2001; Heyes and Huber, 2000), but others, including Cosmides and Tooby, see important distinctions between the various schools. Moreover, many evolutionary anthropologists, human behavioural ecologists, and human sociobiologists have been at pains to differentiate themselves from evolutionary psychology and recognize major theoretical and methodological distinctions between the approaches (Smith *et al.*, 2000).

In this chapter we will focus our attention primarily on research in line with the narrower conception of *evolutionary psychology* as defined by Cosmides and Tooby, because it remains the dominant school of thought within the field, and the broader usage is more diffuse and difficult to characterize. Nonetheless, it is important to bear in mind that a significant number of researchers describing themselves as evolutionary psychologists take issue with aspects of this version, some see no important divisions between the various schools of thought, and some utilize methods and lines of reasoning that we describe as sociobiology, evolutionary anthropology, human behavioural ecology, or the comparative method.

Key concepts

The distinctive theoretical concepts of evolutionary psychology are: first, a focus on evolved psychological mechanisms as the adaptations that underlie human behaviour; secondly, the use of the concept of 'environment of evolutionary adaptedness' (EEA) to reconstruct the adaptive problems faced by our ancestors; and thirdly, an emphasis on *domain-specific* mental organs or modules as evolved solutions to ancestral problems. In this section, we describe each of these concepts in greater depth, and then go on to depict the methodology of evolutionary psychology.

Evolved psychological mechanisms

According to Cosmides and Tooby (1987, p. 281): 'natural selection cannot select for behavior per se; it can only select for mechanisms that produce behavior.' A psychological mechanism is the term they gave to such mental adaptations, the information processing circuits in our brains that shape behaviour. For other researchers (for example, Buss, 1999), psychological mechanisms are defined more broadly to include context-specific emotions, preferences, and proclivities. Psychological mechanisms are assumed to exist in the form that they do because they recurrently solved a specific problem of survival or reproduction over evolutionary history.

Jealousy is provided as an example (Buss, 1994). In ancestral environments, males that experienced jealous emotions when they observed their partner behaving in an overly friendly manner to a rival male, and as a consequence were spurred into action, may have had a selective advantage over males who were indifferent about the possibilities of being jettisoned or cuckolded. How each male went about addressing this problem would depend on factors such as his size, the size of the rival, his personality, and so on. Some males might respond with threats or aggression towards the other male, others with signs of displeasure towards their partner, others with increased vigilance, and others by seeking out a more faithful female. At the behavioural level, it is difficult to predict how an individual will respond to such situations and there is no straightforward answer as to which behavioural strategy maximizes fitness.

However, evolutionary psychologists predict with some confidence that individuals placed in such situations will experience jealous emotions, albeit with varying degrees of intensity, so at the psychological level there is a reliable pattern to be found. Other phenomena proposed as psychological mechanisms include a fear of snakes and spiders, a preference for savannah landscapes, a capacity to learn a spoken language, preferences for particular characteristics in a partner, and a sensitivity to cheating.

Psychological mechanisms are assumed to be complex adaptations that evolved slowly and hence that are unlikely to have undergone any significant change since the Pleistocene. In many respects, they are similar to Lumsden and Wilson's (1981) *epigenetic rules* and Hinde's (1987) *predispositions*, although in some cases the cognitive procedures are specified in more detail. While there is no logical or biologically necessary connection between 'innateness' and modularity, psychological mechanisms are often described as 'innate' or as 'instincts'. For instance, Pinker (1994) describes a 'language instinct', a psychological mechanism that predisposes us to speak complex, fluent grammatical language:

> some cognitive scientists have described language as a
> psychological faculty, a mental organ, a neural system, and
> a computational module. But I prefer the admittedly
> quaint term 'instinct'. It conveys the idea that people know
> how to talk in more or less the sense that spiders know how
> to spin webs. (Pinker, 1994, p. 18)

The use of such terms is unfortunate because they are slippery and vague. Bateson points out that:

> the word 'innate' has at least six separate meanings:
> namely, present at birth; a behavioural difference caused by

a genetic difference; adapted over the course of evolution; unchanging throughout development; shared by all members of a species; and not learned. (1996, p. 2)

Researchers rarely state which meaning is being adopted. What is worse, they may take evidence for one of these meanings as justifying the use of another (Bateson and Martin, 1999).

According to Buss (1999), evolved psychological mechanisms provide non-arbitrary criteria for 'carving the mind at its joints' (p. 52), although the critics of evolutionary psychology question whether such criteria really are non-arbitrary (Lewontin, personal communication). Buss envisages that the mind possesses hundreds, perhaps even thousands, of such specific evolved psychological mechanisms, which are assumed to be universal (or at least, relatively stable) characteristics of human nature. Anthropologist Donald Brown (1991) has documented some of these human universals. For instance, he reports how all people experience certain emotions and express corresponding facial expressions; all have a spoken language, which all have phonemes, morphemes, and syntax; all societies are structured by statuses and roles, and possess a division of labour; and all possess incest avoidance regulations. Humans also possess universals of behavioural development (Bateson and Martin, 1999). With few exceptions, all humans pass the same developmental milestones as they grow up, with most children starting to walk at 18 months, to talk at 2 years, and most reach sexual maturity by their late teens. For evolutionary psychologists, the promise of the evolutionary perspective lies in its power to assist in the discovery, inventory, and analysis of the psychological mechanisms that underpin human nature.

The environment of evolutionary adaptedness (EEA)

The concept of the environment of evolutionary adaptedness was initially developed by the British psychiatrist John Bowlby (1969), influenced by Robert Hinde, to explain why young children the world over develop a strong attachment to their mothers, and why separation can result in extreme distress, including psychiatric disorder. Bowlby argued that the overt attachment of young to their parents should not be regarded as an illness or as dysfunctional behaviour, but rather as an adaptation that in our evolutionary past greatly enhanced the survival prospects of infants. Bowlby asserted that people have lived in modern societies with agriculture, high population density, and complex social institutions for only a few thousand years, while their predecessors lived in small foraging societies for a much longer period of time. The modern world is very different from that experienced by our genus for most of its two million-year history. While attachment and separation anxiety are not necessarily of survival value in contemporary environments, Bowlby envisaged that they were of value at the time and in the environment in which they evolved. The environment of evolutionary adaptedness (EEA) is the term Bowlby gave to this past selective environment. Prior to the late 1960s, there was much confusion over the use of the term 'adaptation' (Gould and Vrba, 1982), and Bowlby's point that evolved characters may be adaptations to past environments was of considerable value.

In their writings on evolutionary psychology, Cosmides and Tooby rapidly adopted Bowlby's notion of the EEA. They also stressed how history and modern culture can change extremely quickly compared to biological evolution, leaving our evolved psychological mechanisms lagging behind:

The recognition that adaptive specializations have been
shaped by the statistical features of ancestral environments
is especially important in the study of human behavior...
Human psychological mechanisms should be adapted to
those environments, not necessarily to the twentieth-
century industrialized world. (1987, pp. 280–1)

Cosmides and Tooby reasoned that if they could establish
what kind of problems our Stone-Age ancestors faced, they
might be able to predict the kind of psychological mechanisms necessary to solve these problems, and hence which
may be expected to have evolved.

Domain specificity

Most evolutionary psychologists believe that minds are
composed of a large number of psychological mechanisms
dedicated to finding quick and efficient solutions to particular problems that were of significance to our ancestors.
One feature of these psychological mechanisms is that each
is believed to have evolved to operate in a specific domain.
Such domains include language, mate choice, sexual
behaviour, parenting, friendship, resource accrual, disease
avoidance, predator avoidance, and social exchange. In contrast, some (although by no means all) non-evolutionary
psychologists may assume that the human mind is a general-
purpose computer with processes that operate across
several domains. Evolutionary psychologists have argued
that from an evolutionary point of view this is highly
implausible. According to Buss (1999), evolved psychological mechanisms tend to be problem-specific because:

(1) general solutions fail to guide the organism to the
correct adaptive solutions; (2) even if they do work, general

solutions lead to too many errors and thus are costly to the organism; and (3) what constitutes a 'successful solution' differs from problem to problem. (p. 52)

Instead, humans should have evolved specialized learning mechanisms that sort experience into adaptively meaningful channels that focus attention, organize perception and memory, and call up specialized procedural knowledge that will generate appropriate inferences, judgements, and choices given the context (Cosmides and Tooby, 1987). In this respect, the mind is described as being like a 'Swiss army knife', with each psychological mechanism analogous to a single blade.

In making the argument that psychological mechanisms are domain-specific, evolutionary psychologists frequently refer to evidence that animals are predisposed to learn some things and not others. A series of elegant experiments by the Berkeley psychologist John Garcia demonstrated that what animals learn varies adaptively across species (Garcia and Koelling, 1966). Garcia gave rats food and then, sometimes after several hours, he gave them a dose of radiation that made them sick. He found that the rats tended subsequently to avoid the food, and they did so because they had learned, often after just a single trial, that food with that particular taste leads to illness. However, the rats struggled to learn an association between the other characteristics of the food and feeling sick, and were extremely slow to learn that a buzzer sound or light predicts illness. From an evolutionary perspective, this makes a lot of sense, as sickness generally results from eating rather than from noises or lights, and taste is a reliable indicator of a food's nature. Garcia's experiments suggested that animals, humans included, were *prepared* by evolution to learn some things more easily and quickly than others.

The methods of evolutionary psychology

Tooby and Cosmides (1989) outline the steps that researchers must go through to do evolutionary psychology.

1. Use evolutionary theory as a starting-point to develop models of adaptive problems the human psyche had to solve.

2. Attempt to determine how these adaptive problems manifested themselves in Pleistocene conditions, and endeavour to establish the selection pressures.

3. Catalogue the specific information processing problems that must be solved if the adaptive function is to be accomplished. Develop a computational theory.

4. Use the computational theory to determine the design features that any cognitive program capable of solving the problem must have, and develop models of the cognitive programme structure.

5. Eliminate alternate candidate models with experiments and field observation.

6. Compare the model against the patterns of manifest behaviour that are produced by modern conditions.

To discourage 'just-so' story-telling, Tooby and Cosmides (1989, p. 41) state

> The desire to leapfrog directly from step one to step six must be resisted if evolutionary biology is to have any enduring impact on the social sciences.

For illustration, consider the example of altruistic behaviour presented by Tooby and Cosmides (1989). The first step is to look to evolutionary theory, where Hamilton's (1964) inclusive fitness theory predicts that individuals ought to be more likely to behave altruistically to close kin.

The second step requires knowledge of our ancestors' selective environment: cooperative exchanges between closely related members of a foraging band might have been critical for survival amongst our Pleistocene ancestors. The third step leads to the reasoning that, for humans to confer benefits on kin, they required cognitive programmes that allow them to determine what are reliable cues indicating relatives and how closely related is a particular individual. As a consequence, the fourth step leads to the conclusion that humans must have psychological mechanisms that allow them to extract this information, and decision rules that use this information to recognize kin. The fifth and sixth steps might, for instance, involve devising experiments that test whether individuals can recognize kin and how they do so, or investigating how people act towards kin and non-kin across different societies.

Buss (1999) outlines two strategies for generating and testing evolutionary hypotheses: a theory-driven strategy similar to the approach of Tooby and Cosmides and an observation-driven strategy. This second approach requires individuals to develop a hypothesis about adaptive function based on a known observation, and to test further predictions based on the hypothesis. Pinker (1997) describes this latter method as 'reverse-engineering', as it starts with the end-product and attempts to reconstruct the steps that led to this point. Other evolutionary psychologists embrace a broader range of methods (Daly and Wilson, 1999). Indeed, in the following section we will describe prominent case studies that test evolutionary hypotheses using psychological experiments, questionnaires, and through analysis of published data records.

Case studies

Here we present three case studies that illustrate the evolutionary psychology approach. We first describe experimental evidence of a psychological mechanism for detecting cheaters. We then examine a study on human mating preferences, and finally look at an evolutionary analysis of homicide.

Psychological mechanisms for detecting cheats

If reciprocal altruism has been important in our evolutionary past, then evolutionary psychologists reason that humans should possess psychological mechanisms that render them sensitive to detecting cheats; that is, individuals that take the benefits from a social exchange without paying the costs.

Statements such as 'If you take the benefit, then you must pay the cost' are known as conditional rules. They can be represented in abstract terms as 'If P, then Q'. One widely used experimental paradigm for exploring people's ability to detect violations of conditional rules has been the Wason selection task. Psychologist Peter Wason (1966) wanted to know whether people think by testing hypotheses and devised an experiment to determine whether they were good at detecting violations of conditional rules. He found that people reason logically only in restricted contexts and that the subject matter people are asked to think about seems to affect how well they do on these tests. Consider the task to detect violations of the abstract rule 'If a person has a 'd' rating, then the documents must be marked code 3' depicted in Figure 5.1a. Wason found that typically less than 25% of people answer this task correctly. You can try this test for yourself before reading what the correct answer should be.

167

a. Abstract problem

Part of your new clerical job at the local high school is to make sure that student documents have been processed correctly. Your job is to make sure the documents conform to the following rule:

If a person has a D rating, then the documents must be marked code 3

You suspect the secretary you replaced did not categorize the student's documents correctly. The cards below have information about the documents of four people who are enrolled at this high school. Each card represents one person. One side of a card tells a person's letter rating and the other side of the card tells that person's number code. Indicate only those card(s) you definitely need to turn over to see if the documents of any of these people violate this rule.

b. Drinking age problem

In a crackdown against drunk drivers, Massachusetts law enforcement officials are revoking liquor licenses left and right. You are a bouncer in a Boston bar, and you'll lose your job unless you enforce the following law:

If a person is drinking beer, then he or she must be over 21 years old

The cards below have information about four people sitting at a table in your bar. Each card represents one person. One side of a card tells what a person is drinking and the other side of the card tells that person's age. Indicate only those card(s) you definitely need to turn over to see if any of these people are breaking this law.

Adapted from Figure 3.3 and examples given in Cosmides & Tooby (1992).

Figure 5.1

Most people presented with this abstract problem selected only the D card, or the D and 3 cards, as necessary to check for violations. In fact, the right answer is to turn over the D and 7 cards. This is because, to establish that every D card has a 3 on the flip side it is clearly necessary to turn over the D card, but also important to establish that the 7 is not a D. Whether the 3 is a D or not is irrelevant, as the rule does not insist that D is the only rating with the code 3. Now compare your performance with the task shown in Figure 5.1b. Surprisingly, despite the fact that the drinking age task depicted there is logically exactly the same, people consistently perform better on this task, with approximately 75% of subjects giving the logically correct response of 'drinking beer' and '16 years of age'. In both tasks, individuals are given a conditional rule of the form If P then Q (i.e. if D then 3, or if beer then over 21), and asked what they need to do to determine whether this rule has been violated. The rule is violated only when P is true but Q is false, and thus in both cases the answer is to check P (the D or beer card) and not Q (the 7- or 16-years-old card).

Such experiments suggest that human reasoning changes depending on the subject matter about which one is reasoning, but prior to an investigation by Leda Cosmides there was no satisfactory theory that could account for these content effects. As part of her doctoral dissertation at Harvard University, Cosmides set out to establish whether the contexts in which people reason logically made sense in evolutionary terms. In particular, she was interested in the hypothesis that a history of reciprocal altruism among our ancestors would have fashioned us with a cheater detection mechanism that biased our reasoning.

In an elegant series of experiments that expanded Wason's findings, and for which she was awarded the AAAS Behavioral Science Research Prize, Cosmides found that when subjects are asked to look for violations of conditional rules that express social contracts their performance improves dramatically (Cosmides, 1989; Cosmides and Tooby, 1992). According to Cosmides, the reason most people get the abstract problem wrong but the drinking age task correct is that only in the latter case does logic coincide with cheater detection (Cosmides and Tooby, 1992). The drinking age task has a content equivalent to 'If you take the benefit, then you pay the cost'. Here drinking beer is the benefit, being over 21 is the cost, and drinking alcohol under age is cheating by violating a social norm. Cosmides's experiments ruled out alternative explanations, such as that performance was better on some tasks than others because the content was more familiar. Even when subjects were given an entirely unfamiliar rule, such as 'If a man eats cassava root, then he must have a tattoo on his face', they responded with a high level of success provided the preamble gave them sufficient information to establish that the rule was a social contract. Most compelling of all, Cosmides was able to switch the order of the rules so that the logically correct answer conflicted with the social contract theory, and subjects responded in a manner consistent with the cheater detection hypothesis (for a description of these experiments see Cosmides and Tooby, 1992).

Cosmides and Tooby argue that people are tuned to attend to situations in which people take the benefit without paying the cost. Although not all psychologists accept Cosmides and Tooby's interpretation of these findings, few would dispute that Cosmides's experiments have reinvigorated this area of research and made a valuable contribution to the field. It remains an intriguing and highly plausible possibility that our minds are equipped with cognitive adaptations for social exchange, of which one procedure is a psychological mechanism dedicated to looking for cheats.

Sex differences in mate choice

As natural selection operates through the differential reproduction of individuals, any psychological mechanisms that guide reproduction should be especially strong targets of selection. As a consequence, courtship and sex have been a principal focus of evolutionary psychology (Buss, 1994). Indeed, the great bulk of research in evolutionary psychology has been focused on human mating behaviour. One question that has received considerable attention is whether evolution has fashioned us with preferences for particular characteristics in the opposite sex that influence our choice of mating partners.

Trivers (1972) proposed that females should seek to mate with males who show the ability and willingness to invest resources connected with parenting such as food, shelter, territory, and protection. There is now considerable experimental evidence from studies of animals that females frequently best maximize their reproductive success by prioritizing gaining access to resources. Perhaps humans are no different in this regard. Evolutionary psychologists have reasoned that, from the perspective of our ancestors in the EEA, women faced the burdens of internal fertilization, a nine-month gestation, and lactation, and consequently would have benefited by selecting mates who possessed and were willing to provide such resources (Buss, 1994). They also suggested that females might be selected to favour males that display cues indicating their wealth, such as status, or their potential to accrue substantive resources in the future, such as intelligence, hard work, and ambition.

In contrast, in most mammals male parental investment is small compared with that of females, and hence males can most effectively maximize their reproductive success by prioritizing mating with many females and by choosing females that are fertile. Evolutionary psychologists argue that thousands of generations of selection have favoured the evolution of psychological mechanisms in males that render the prospect of many sexual partners desirable and females of high fertility attractive (Buss, 1994). As human female fertility is highest in the early twenties, men are predicted to prefer younger to older women. Some researchers have suggested that standards of beauty reflect an evolved preference for physical traits that are generally associated with youth, such as smooth skin, good muscle tone, and an optimal waist-to-hip ratio.

To test these hypotheses, psychologist David Buss, currently at the University of Texas, Austin, carried out an extensive series of cross-cultural studies to determine whether human mate choice shows consistent patterns the world over (summarized in Buss, 1994). One investigation involved Buss and his collaborators interviewing over ten thousand people in thirty-seven different cultures (Buss *et al.*, 1990). On the basis of these analyses, Buss concluded that there is a broad cross-cultural consensus about what attributes are important in a mate, and that the sexes show the distinct patterns predicted by evolutionary psychology reasoning. For instance, Buss found that:

> Women across all continents, all political systems
> (including socialism and communism), all racial groups,
> all religious groups, and all systems of mating (from
> intense polygyny to presumptive monogamy) place
> more value than men on good financial prospects.
> (1994, p. 25)

In contrast, men typically placed more value than women on the physical attractiveness of their partner:

> Men worldwide want physically attractive, young, and sexually loyal wives who will remain faithful to them until death. These preferences cannot be attributed to Western culture, to capitalism, to white Anglo-Saxon bigotry, to the media, or to incessant brainwashing by advertisers. (1994, p. 70)

Buss also uncovered clues suggesting an evolutionary past that favoured men that had short-term mating in their sexual repertoire:

> sexual fantasy ... lust, the inclination to seek intercourse rapidly, the relaxation of standards, shifts in judgements of attractiveness, homosexual proclivities, prostitution, and incestuous tendencies are all psychological cues that betray men's strategies for casual sex. (1994, p. 85)

However, we suggest that these findings need to be kept in perspective. Buss's study found that mutual attraction, dependable character, emotional stability, and a pleasing disposition were the four traits deemed most important to mate choice by both sexes. Good financial prospects was on average rated only the twelfth most important factor influencing mate choice in females, and good looks were rated only tenth by males. Moreover, Buss found that, for most traits, knowing where a person lives tells you more about what he or she values in a mate than knowing the person's gender, indicating that sex differences are comparatively unimportant compared with cross-cultural differences. For instance,

> The trend for men to value chastity more than women holds up worldwide, but cultures vary tremendously in the value placed on chastity. At one extreme, people in China, India, Indonesia, Iran, Taiwan, and the Palestinian Arab

areas of Israel attach a high value to chastity in a potential mate. At the opposite extreme, people in Sweden, Norway, Finland, the Netherlands, West Germany, and France believe that virginity is largely irrelevant or unimportant in a potential mate. (Buss, 1994, p. 68)

In addition, the criteria on which standards of attractiveness are judged vary greatly from one culture to the next, with some cultures, for instance, preferring plump to slim builds, and others preferring dark to light skin colour. Moreover, much of the research in this area is carried out by giving questionnaires to university and college students, and one might question to what extent students in different countries really represent distinct cultures. It would be interesting to find out whether the reported sex differences remain if the studies were carried out on groups such as the Hadza (Tanzania), Ache (Paraguay) or Mapuche (Chile). The reliability of questionnaires and self-reports has also been queried (Nisbett and Wilson, 1977; Aunger, 1994), which may be a particularly acute problem in studies of sexual behaviour. Nonetheless, Buss's analyses provide some of the broadest evidence to date that evolved psychological mechanisms may be universal features.

Homicide

All around the world the folk literatures of distinct cultures abound with Cinderella stories involving a cruel or evil step-parent. For Martin Daly and Margo Wilson, two psychologists at McMaster University in Canada, the ubiquity of these stories reflects a genuine, dark, and disturbing aspect of human societies. Daly and Wilson have used an evolutionary psychology perspective to inform a study of homicide, leading to a number of novel questions, hypotheses, and conclusions. In fact, it was in the flyers to Daly and Wilson's (1988) pioneering book *Homicide* that the phrase 'evolutionary psychology' was first coined. A clear prediction made from Daly and Wilson's evolutionary perspective was that, as they are unrelated, substitute parents will generally tend to care less for children than natural parents, with the result that children reared by people other than their natural parents will more often be at risk. Raising a child involves considerable costs and substitute parents may be less likely than natural parents to experience the emotional rewards that make the costs of parenthood tolerable.

In an extensive analysis of data on infanticide in Canada and the United States, Daly and Wilson documented the fact that there was a very real and substantially elevated risk to children residing with one natural parent and one step-parent. For instance, the American Humane Association detected 279 fatal incidences of child abuse in 1976, of which 43% dwelt with a substitute parent, considerably more than would be expected by chance. Another survey of child abuse in Canada in 1983 gave a similar pattern of results. Daly and Wilson argued that poverty, which is also associated with child abuse, does not explain the association between abuse and step-parenthood. According to Daly and Wilson, the more common social science explanation for the difficulties encountered in step-relationships is that these difficulties are in fact caused by the 'myth of the cruel step-parent' and by the fears of the child. The evolutionary psychology view appears to present a more compelling description of the observed patterns of behaviour.

Daly and Wilson also used their evolutionary perspective to investigate adult murders outside of the family. In a 10-year survey of Canadian homicide, they found that the predominant form of murder involved men killing unrelated men. In fact, in accounting for single sex murders among adults, Daly and Wilson recorded 2861 male–male cases to 84 female–female cases, showing the former to be 34 times as frequent as the latter. A survey of 35 studies of homicide from around the world revealed that this difference between the sexes is found in every single population in which it has been investigated. According to Daly and Wilson, there is no known human society in which the level of lethal violence among women even begins to approach that among men.

Why should there be a universal sex difference in homicidal aggression amongst humans? Daly and Wilson explain how evolutionary biology provides an answer. Trivers (1972) argued that across all sexual species, the sex that makes the greater parental investment tends to become the crucial resource limiting the fitness of individuals of the less investing sex, so that selection favours competition among the latter for access to mates. In humans, females are the sex making the greater investment in raising offspring and males could father many children if they had access to multiple mates, potentially many more than an equivalent female. There is strong evidence that the selective history of our ancestors was one that involved mild but sustained polygyny; in fact, such is the norm in many human societies today. While females are likely to have been competing among each other for quality males too, the variance in male fitness was probably greater than the variance in female fitness. In other words, the successful males are big winners with many wives and offspring, and the losers may do extremely poorly, while virtually all females have some intermediary level of reproductive success. From this evolutionary perspective, where there are big rewards for competition between males for access to females, the entire life history of males may favour higher risk strategies. The more intense the competition, the more likely it becomes that selection will favour psychological mechanisms in males rendering them prone to risky competitive tactics, including escalated fighting even to the point of death. Daly and Wilson (1983) showed that this hypothesis is supported by related studies of risky behaviour in humans. For instance, they pointed out that males are more prone to dangerous driving and suffer elevated rates of mortality on the roads. Another example is that 93% of robberies and 94% of burglaries in the United States in 1980 were perpetrated by males. Males are not poorer than females but they would seem to be more prone to taking risks. Daly and Wilson hypothesize that the risks that males take may reflect a past history of selection that has fashioned their minds for competition.

Critical evaluation

Much of the criticism levelled at evolutionary psychology is identical to that directed at sociobiology; indeed, many critics see no meaningful distinction between these two schools (e.g. Rose and Rose, 2000). Rather than repeat ourselves, we refer the reader back to the penultimate section of Chapter 3, where we discuss these charges. To reiterate briefly, allegations of genetic determinism or prejudice on the part of leading sociobiologists or evolutionary psychologists are usually unfounded; charges of reductionism are misguided; however, criticism on the grounds of 'Just so' evolutionary story-telling and a superficial reading of the relevant literature are frequently justified. Here we concentrate on evaluating the distinctive characteristics of evolutionary psychology, focusing on issues related to the environment of evolutionary adaptedness, domain specificity, and their general evolutionary perspective.

Evaluating the concept of the EEA

Early work by evolutionary psychologists asserted that the human mind was fashioned over the last two million years for a past world of hunting and gathering on the African plains of the Pleistocene. For instance, Cosmides and Tooby wrote:

> Our species spent over 99% of its evolutionary history as hunter–gatherers in Pleistocene environments. (1987, pp. 280–1)

Daly and Wilson (1999) point out that much of the dissatisfaction with the EEA concept has derived from an equation of the EEA with a stereotype of a Pleistocene African savannah. Cosmides and Tooby have informed us that they never adhered to this stereotype, and that their early writings on the EEA were simplified to reach an 'evolutionarily-naive' audience that tended to regard all human behaviour to be of utility in current environments. Unfortunately, a damaging EEA-as-Pleistocene-African-savannah stereotype pervades the evolutionary psychology literature.

What is wrong with the notion of the human EEA as a particular time and place? The problem is that comparatively little is known about the lifestyle of our ancestors throughout the Pleistocene. Consequently, the EEA concept has engendered a wealth of undisciplined speculation and story-telling in which virtually any attribute can be regarded as an adaptation to a bygone Stone-Age world. A stereotypical notion of the EEA implies that the Pleistocene hunter–gatherers exhibit little variability in time or space,

which a number of researchers have pointed out is false
when one considers that Stone-Age peoples lived not only
on the African savannah, but in deserts, next to rivers, by
oceans, in forests, and in the Arctic (Foley, 1996; Boyd and
Silk, 1997). The evolutionary psychology literature makes
common reference to the observation that 'humans spent
99 per cent of their evolutionary history as hunter–
gatherers'. Yet every human descends from ancestors collectively subject to natural selection for three and a half billion
years, which leaves the '99 per cent' figure arbitrary.

Neither is a description of our ancestors as 'hunter–
gatherers' a sufficient account of their life history to be able
to reconstruct the relevant selection pressures. Wasps, rats,
and blue tits are all hunter–gatherers in the sense that they
both hunt live prey and gather other foods. Of course, they
do not exhibit the cooperative, coordinated, socially organized, linguistically guided hunting and gathering that
modern human hunter–gatherers exhibit, but the point is
that it is not known whether our ancestors during the
Pleistocene did so either (Foley, 1996). Many authoritative
archaeologists and anthropologists believe that *Homo erec-
tus* and even Neanderthals lived completely different lives to
modern hunter–gatherers. To what extent they had sophisticated linguistic abilities, hunted large game, shared food,
and had home bases, for example, is open to dispute. If, as
many believe, these characteristics emerged as late as the
upper Paleolithic, around forty thousand years ago, any
focus on the earlier Pleistocene would be misplaced.

More recently, Tooby and Cosmides have clarified their
position:

> [The EEA concept does not refer to a single] place or
> habitat, or even a time period. Rather, it is a statistical
> composite of the adaptation relevant properties of the
> ancestral environments encountered by members of

ancestral populations, weighted by their frequency and their fitness consequences. (Tooby and Cosmides, 1990a, pp. 386–7)

However, this conceptualization may be problematic in a different sense. Can the 'new' EEA concept be put to use, in the manner that Tooby and Cosmides (1989) originally claimed, to develop models of adaptive problems the human psyche had to solve? How could one compute a 'statistical composite' of all the relevant environments encountered by our ancestors, and weight them accordingly? Comparative analyses of animal abilities suggest that many human behavioural and psychological traits have a long history. Some human behavioural adaptations, such as maternal care or a capacity to learn, may even have evolved in our invertebrate ancestors. Many perceptual preferences will be phylogenetically ancient. For example, an understanding of causal relationships may be common to mammals and birds. Much social behaviour, such as forming stable social bonds, developing dominance hierarchies, an understanding of third-party social relationships, and coordinated hunting, probably evolved in our pre-hominid primate ancestors. A capacity for true imitation may also have evolved in pre-hominid apes. Yet if researchers are going to use the EEA as Cosmides and Tooby originally outlined, they need to identify a particular time period and class of ancestor when the relevant psychological mechanisms evolved, and then weight that and all subsequent environments accordingly. In principle, EEA supporters could carry out a phylogenetic analysis to determine the earliest known ancestor exhibiting a trait. In practice, this is never done and, as little is likely to be known about that particular ancestor and most of its descendants, it would be an extremely time consuming exercise that would generate only vague speculation.

Perhaps the real virtue of the EEA concept is more modest. The EEA encourages researchers to recognize that humans, like all species, exhibit *some* adaptations to past environments that are not necessarily of current utility. The originator of the EEA concept, John Bowlby, was concerned with the mother–child relationship, which we might envisage has a degree of constancy across environments and over time. There is a strong argument that the EEA concept was important in developing an understanding of childhood separation anxiety and attachment (Hinde, 1987). Similarly, researchers do not need to know the precise conditions in which humans evolved to make the reasonable guess that salts and sugars may not have been in abundant supply so that their reinforcing properties may not have been counterbalanced by regulatory processes operating against consuming excess (Bateson and Martin, 1999). The question is what proportion of human behavioural traits can be assumed to have evolved in all relevant past environments?

In conversation with us, John Tooby suggested that one doesn't need to know when traits *first* evolved to use the EEA concept, as the behavioural regulatory machinery would have been modified by selection up until the Pleistocene. This line of reasoning brings researchers back to the position where knowledge of stone-age conditions is all that is needed to reconstruct the selective environment of our ancestors. However, this argument is based upon a number of assumptions, including that there was genetic variation in psychological traits up until the Pleistocene, that no significant mental structure carries any historical legacy of selection prior to the Pleistocene, that there has been no meaningful selection on psychological mechanisms since the Pleistocene, and that evolutionary change occurs at a particular rate. These assumptions are not in themselves unreasonable, but they remain highly disputed. Another caveat for the EEA argument is that, at best, it can only be partly true. Human beings cannot be exclusively adapted to a past world and not at all adapted to modern life, otherwise we would not be able to exist. It would be puzzling if our ancestors really started to thrive as soon as they left their EEA, yet it is in the Holocene, the period since the Pleistocene, that we see the explosion in human numbers and human colonization of the globe. This population growth suggests that a significant fraction of human characteristics remain adaptive even in modern environments which share features with those of our ancestors. Any assumption that natural selection on humans has stopped, that no genetic variation underlies human psychological characters, and that measuring human fitness is a waste of time, is questionable. This is well illustrated by a study by Pawlowski, Dunbar, and Lipowicz (2000) which demonstrated that taller men are reproductively more successful than shorter men, suggesting that, in contemporary populations, there is active selection for stature in male partners, perhaps brought about through female preference or competition amongst males. This study shows that, even in the modern world, with widespread use of contraception and extensive medical care, natural selection is still in operation.

Moreover, the view that modern human populations are adapted to an ancestral Pleistocene habitat is misleading because it portrays humans as passive victims of selection rather than as potent constructors of their niche. It is a distortion to regard evolution as a process by which organisms solve problems set by the environment (Lewontin, 1983a). Niche-construction theory represents one increasingly accepted strand of evolutionary genetics that lays emphasis on the fact that organisms themselves *modify* important components of their selective environments (Odling-Smee *et al.*, 1996; Laland *et al.*, 1996; 2000). For humans, our capacity to create solutions continuously to self-imposed problems reflects the fact that we are very adaptable creatures. Moreover, to a degree that surpasses other species, human mental processes must contend with a constantly changing information environment of their own creation (Flinn, 1997). The flexible nature of our learning and culture allows us to survive and flourish in a broad range of settings. This adaptability means that, rather than being adapted to a particular environment, humans adapted to a broad range of environments that they and their ancestors were involved in constructing.

Psychological traits may be domain-general

One contentious aspect of evolutionary psychology is the stress laid on domain-specific psychological modules. Many researchers believe that evolutionary psychologists have overplayed the modularity of the human brain, and maintain that minds have many domain-general features. Cosmides and Tooby (1987) characterize the difference between the standard social science view and their perspective as representing a choice between two models of the mind, one that lays emphasis on a small number of domain-general processes versus another stressing a large number of domain-specific modules. However, domain-general and domain-specific represent poles of a continuum. Evolutionary psychologists are surely correct to point out that there are efficiency benefits to be gained by mental division of labour and that at times evolution would favour specialization of psychological processing. Yet one can also have too much specificity. It would simply not be feasible to construct a brain that allocates a specific psychological module to every conceivable event an individual might encounter, as the costs in terms of neural circuitry and information processing would be huge. There is no intrinsic virtue to mental specificity; general solutions will be favoured when they can do a good enough job at low cost. For example, human beings may have a psychological module that leaves them predisposed to fear snakes, but they do not have modules that discriminate between dangerous and harmless snakes, or constricting and poisonous species, despite the fact that one can envisage some utility to such discriminations. Domain-general processes are no more incompatible with evolutionary theory than domain-specific processes.

Garcia's experiments are frequently hailed by evolutionary psychologists as demonstrating the gene-biased nature of classical conditioning in particular, and more generally the inadequacy of associative learning theory (the idea that we learn by forming associations between events). Yet associative learning is widespread and has general properties that allow animals to learn about the causal relationships among a wide variety of events (Mackintosh, 1974; Dickinson, 1980). Learning can occur via quite simple rules; for example, one theory known as the Rescorla–Wagner rule (1972) has proved useful in explaining the results of experiments on foraging in honey bees, avoidance conditioning in goldfish, and inferential reasoning in humans. Even some of the most enthusiastic supporters of a modular view of the brain (e.g. Shettleworth, 2000) accept that, while what is learned may vary adaptively across species, how it is learned does not. Natural selection may have fashioned us to be prepared to form some associations more readily than others, and built in some motivational priorities, but many psychologists regard this as more tinkering with the general system than constructing an independent set of species-specific learning processes (Bolhuis and MacPhail, 2001).

Cosmides and Tooby (1987) have argued that learning should not be regarded as an alternative to evolutionary explanations. However, our capacity to learn is an unusual adaptation. It has a property that makes it different from other adaptive responses of phenotypes to the environment, such as calluses on the hands (Buss, 1995); namely, that it is an information gaining subsystem. Its function is to acquire and store information about the world, information that will generally guide behaviour towards adaptive goals but information that nonetheless could not be specified in our genes. Rather than fashioning us with brains hardwired to recognize apples as food and sand as not food, natural selection has given us a flexible information gaining problem solver, with instructions to seek food when blood sugar levels are low and to recognize apples as food because they taste good while sand doesn't. A rule like 'Actions that are followed by a positive outcome are likely to be repeated, while those followed by a negative outcome will be eliminated' is domain-general in the sense that it can be equally applied to behaviour concerned with finding food, avoiding predators, or seeking a mate. This particular rule was first described by American psychologist Edward Thorndike in 1911, and is known as 'The Law of Effect'. While comparative psychologists still argue over the details and rarely specify the problem in informational terms, few would dispute that something approximating this rule governs much human learning. If researchers want to know why individuals prefer eating apples to sand, the best explanation is an evolutionary one, as our learning about foods is constrained to substances of nutritional value. However, if researchers want to know why some humans eat apples and others snails or curry, arguments based on biological evolution have comparatively little to offer. This is not to say that specialized processes play no part in learning. We may well be predisposed to adopt the behaviour of the majority, imitate the successful or experience norm violations as aversive, for instance. However, our genes specify a tolerance space for our acquired information but rarely the details within it.

Much of the debate over the merits of evolutionary psychology explanations revolves around the extent to which human developmental processes are under tight genetic regulation in which developmental outcomes are pre-specified and channelled, as opposed to a more flexible system in which pre-specification of regulatory development is minimal. Evolutionary psychologists are content to assume past selection for different properties of mind, such as altruism or jealousy. However, in the absence of any established neurobiological theory of how (or indeed whether) genes that bias the growth and connections of neurons during development influence the relevant psychological states, a fundamental part of the causal pathway is missing. Researchers cannot carry out experiments on humans to establish whether 'altruism' can be subject to selection. To our knowledge, no-one has ever shown that 'jealousy' has a genetic basis, or is heritable. We agree that it is quite plausible that natural selection may have favoured particular psychological states in specific past environmental contexts. However, given the immense developmental plasticity and flexibility of the human brain, it is also conceivable that 'jealousy', 'altruism', and many other psychological states are better regarded not as adaptations but as a by-product of our extraordinary adaptability.

Learning processes are not the only psychological processes to exhibit domain-general properties. The senses are classic examples of modular division of labour, yet share a number of functional properties, such as a sensitivity to contrast, a tendency to habituate, and a tendency to give a bigger response to a bigger stimulus (Shettleworth, 2000). Fodor (1983), a philosopher who pioneered the notion, regarded modularity as operating primarily at the level of these sensory input systems to the brain, with central cognitive processing more general across domains. Sensory inputs feed into some quite general cognitive processes, such as planning, reasoning, mental state attribution, and problem solving. It is even conceivable that cognitive modularity has been reduced during recent human evolution, allowing more integration of information and communication amongst modules (Mithen, 1996). The more extreme evolutionary psychologists appear to regard cognition as modular right through from perception to action, the implication being that modules operate in parallel and rarely interact (Bolhuis and MacPhail, 2001).

When we asked Cosmides and Tooby whether they would accept that many psychological traits are domain-general they responded with an emphatic 'Of course!', and pointed to experimental studies of theirs that had demonstrated as much (for example, Brase *et al.*, 1998). However, a hypermodularized depiction of the mind continues to pervade much of the evolutionary psychology literature (e.g. Buss, 1999).

Adaptationism and evolutionary biology

Most evolutionary psychologists adhere to a branch of evolutionary thinking known as 'adaptationism'. Unfortunately the term 'adaptationism' is used in at least two quite distinct ways by enthusiasts and by critics of this perspective. Adaptationists take inspiration from George Williams' (1966) *Adaptation and Natural Selection*, which advocated a much more rigorous use of the term 'adaptation', and argued that natural selection was a sufficient theory to explain most of what is important about evolution. In spite of this, for their critics, adaptationists are researchers who describe virtually all characters as adaptations and who underestimate the importance of other processes in evolution. While many evolutionary psychologists are commendably disciplined in their attribution of adaptations, which are carefully distinguished from exaptations and by-products (for definitions see Chapter 4), others appear less cautious. Moreover, critics of evolutionary psychology feel that these researchers underestimate the significance of evolutionary processes other than the natural selection of genes (Lloyd and Feldman, 2001). The fact that few evolutionary psychology studies refer to the findings of modern evolutionary biology reinforces the suspicion that evolutionary psychology has become detached from recent developments in evolutionary thinking, which over the last 30 years have increasingly stressed a wide range of processes (Endler, 1986b; Futuyma, 1998). The contemporary reality is that evolution is a much more complex phenomenon than that portrayed in evolutionary psychology textbooks (Lloyd and Feldman, 2001).

Endler (1986b) identified 21 processes that are instrumental in evolutionary change, stressing that his list was incomplete. It has become clear that natural selection operates at several different levels and, unlike 25 years ago, multi-level selection models are now a common and respectable feature of evolutionary genetics. Selfish DNA such as microsatellites, and selfish genes such as transposons and segregation distorters, are examples of selective processes operating below the level of the individual, while above this level an increasing proportion of specialists accept the idea that species selection and clade selection could be important (Stearns, 1986; Lloyd, 1994; Rice, 1995; Sober and Wilson, 1998; see also the articles in Rose and Lauder, 1996). Indeed, few evolutionary psychologists appear to realize that among the converts to the idea of 'clade selection' can be found their guru George Williams (1992), previously renowned for his criticism of group selectionist arguments.

Nor is measurement of fitness straightforward (Lewontin, 1974). Endler (1986a, p. 33) writes, 'there are many different definitions and measures of fitness' and reduces the multitude of terms and methods to a core five concepts. Many evolutionary psychologists characterize Hamilton's inclusive fitness theory as the cornerstone of modern evolutionary thinking (Cosmides and Tooby, 1987; Ketelaar and Ellis, 2000), yet this represents a small subset of models used for special purposes in evolutionary understanding, and which cannot handle sexual selection, multi-locus selection, or multi-level selection (Lloyd and Feldman, 2001).

Identifying what constitutes a character that is subject to natural selection is a well recognized and stubborn problem within contemporary evolutionary biology which has countless difficulties but no universally accepted solution (Wagner, 2001). For instance, it is well known that human evolution is characterized by neoteny, that is a slowing down in development, so that in certain characteristics the anatomy of the adult human being resembles the infant ape more than it resembles the adult ape. Lewontin (2000) points out that there have been many speculations about why natural selection might have favoured a protruding chin in humans, making it an exception to the rule of neoteny. In reality, the evolution of neotenous development has produced smaller jawbones, but the dentary and mandibular bones have receded at different rates, most likely as a consequence of developmental constraints, and the chin is an incidental outcome. In other words, the chin is not a character that has been favoured by natural selection. While Cosmides and Tooby have been admirably cautious in their use of the term adaptation, few evolutionary psychologists take time to ensure that their traits truly are an integrated unit of development selected for a particular function rather than an incidental feature to which a name has been given.

Similar problems relate to identifying adaptations. It is sometimes possible to make an educated guess as to whether a character is an adaptation by drawing inferences about which traits might be expected to have been favoured by selection in the past, based on knowledge of evolutionary processes and ancestral environments (Cosmides and Tooby, 1987; Tooby and Cosmides, 1990a). The likelihood of such inferences being correct is a matter of some controversy. As investigators are rarely completely ignorant of the nature of the character that will eventually be described as an adaptation, they may be in a position to 'cheat' and devise an evolutionary story that predicts qualities of the character that are already known to exist. Under such circumstances, confirmation of the predictions through experiments or questionnaires would hardly be compelling. Researchers rarely restrict the application of this method to characters for which the relevant features of the ancestral environment are reasonably well known, or their predictions to phenomena that are not self-evident. Given the well-documented difficulties of identifying adaptations (Rose and Lauder, 1996), researchers would be well advised not to settle for a single line of evidence. Independent corroboration that the observed character has been correctly identified as an adaptation can be provided through the use of mathematical models, the comparative method, phenotypic manipulations, or by inference from the character's 'engineered' or design properties (Rose and Lauder, 1996; Sinervo and Basolo, 1996; Orzack and Sober, 2001).

There are other respects in which evolutionary psychology appears to circumvent the complexities of evolutionary biology. For instance, Cosmides and Tooby argue that:

> the complex architecture of the human psyche can be
> expected to have assumed approximately modern form
> during the Pleistocene ... and to have undergone only
> minor modifications since then. (1987, p. 34)

This reasoning is based on the assumption that complex characters evolve slowly. However, while it is a reasonable supposition that complex traits will evolve more slowly than simple ones, evolutionary biology has not yet gained a sufficient understanding to be able to pin reliable quantitative estimates on rates of character change. It is not known if complex adaptations always take millions of years to evolve, but the evidence for those traits studied is, if anything, to the contrary. Selection experiments and observations of natural selection in the wild have, over the last 20 years, led to the conclusion that biological evolution may be extremely fast, with significant genetic and phenotypic change sometimes observed in just a handful of generations (e.g. Dwyer *et al.*, 1990; Grant and Grant, 1995; Reznick *et al.*, 1997; Thompson, 1998). Recently, Kingsolver and colleagues (2001) reviewed 63 studies that measured the strength of natural selection in 62 species, including over 2,500 estimates of selection. They concluded that the median selection gradient (a measure of the rate of change of fitness with trait value) was 0.16, which would cause a quantitative trait to change by one standard deviation in just 25 generations. While it is possible that selection gradients may be weaker when measured over larger time scales (Gingerich, 1983), it is clear that substantive biological evolution can occur in thousands of years, or less. A quotation from *Sociobiology: The New Synthesis* remains apt:

> The theory of population genetics and experiments on other organisms show that substantial changes can occur in the span of less than 100 generations [and] it would be false to assume that modern civilizations have been built entirely on capital accumulated during the long haul of the Pleistocene. (Wilson, 1975, p. 569)

Finally, given the prevalence of evolutionary psychological explanations for sex differences in human behaviour and anatomy in terms of sexual selection, it is worth reflecting on the basics that would need to be in place for such hypotheses to be viable. As an example, consider the recent interest engendered by research into human mate choice and character symmetry. Fluctuating asymmetry (FA) is a measure of the symmetry of a bilateral character (e.g. ear length or hand breadth) that fluctuates, it is supposed, in response to internal and external stress factors such as inbreeding or parasitic infection. A high level of FA (e.g. one foot longer than the other) is thought to indicate poor condition, on the assumption that it requires a sound metabolism to grow perfectly symmetrical features. Some models of sexual selection suggest that females choose a male with traits indicating that he is strong and healthy, on the grounds that their offspring will inherit these 'good genes' (Zahavi, 1975), and some researchers have suggested that symmetry (or low FA) represents such a trait (e.g. Møller, 1990).

Several evolutionary psychology studies conclude that women find men with symmetrical features more attractive than their asymmetric counterparts and posit a 'good genes' explanation (an overview of these studies can be found in Cartwright, 2000). Yet consider some of the fundamentals that would have to be established to provide reasonable support for this hypothesis:

(1) There would have to be evidence that there is, or has been, genetic variation underlying female preferences and the symmetry of male faces.

(2) Male facial symmetry and female preferences for symmetrical faces would have to be shown to be (or have been) heritable.

(3) Male facial symmetry and female preferences would have to be shown to co-vary with fitness, or to have co-varied with fitness in the past.

(4) There would have to be evidence that male facial symmetry is, or has been, sexually selected (as opposed to naturally selected).[3]

Not only is this evidence rarely provided, but a number of biological studies have shown that the association between

---

[3] A good illustration of how traits seemingly fashioned by sexual selection can actually be the product of natural selection is provided by Heather Proctor's elegant studies of the mating displays of water mites (Proctor, 1992, 1993). Individuals of both sexes feed on aquatic invertebrates by sitting with their front legs spread out and pouncing on prey items that they detect through vibrations in the water. Males have taken advantage of this pre-existing female response by evolving a sexual display that involves the vibration of their front legs at the same frequency as the prey, and depositing spermatophores when the females grab them. A series of experiments and comparative analyses reveal no evidence that sexual selection has fashioned female mate choice, but considerable support for the sensory exploitation hypothesis. Yet a study that focused solely on sexual behaviour could easily draw the erroneous conclusion that the females are choosing males with 'good genes' or protein-rich spermatophores.

FA and fitness is tenuous and perhaps an artefact of selective reporting, that there are not consistent correlations among different measures of FA on the same organisms, that the human traits commonly used are rarely measured accurately enough to prevent FA from being confounded by measurement error, and that the heritability of FA for most appropriately measured traits is close to zero.[4] While it may be tempting to conclude that collecting data on human reproductive success and heritability would be ineffectual in a modern world where fitness is clouded by use of contraception, and where environments are very different from those of our ancestors, other studies have found strong evidence for ongoing selection in contemporary human populations and demonstrated the feasibility of testing these assumptions in humans (Durham, 1991; Pawlowski *et al.*, 2000; Smith *et al.*, 2000).

If evolution is a complex multi-faceted phenomenon, if many evolutionary processes, including drift and mutation, are operating at the same time, if evolutionary history is important, if selection is operating at different levels, and if evolutionary rates can sometimes be fast, it makes the business of predicting and interpreting psychological adap-

---

[4] Markow (1995) and Clarke (1995) both conclude that any association between FA and fitness is tenuous, speculative, and character-specific. A meta-analysis by Palmer (2000) concludes that relations between FA and individual attractiveness or fitness may be a result of selective reporting. Schlichting and Pigliucci (1998) cite studies that find no consistent correlations among different measures of stability on the same organism. Palmer and Strobeck (1997) provide a good discussion of the confounding effects of measurement error. In criticizing Møller and Thornhill's (1997) selection of studies to generate a estimate of the heritability of FA, Leamy (1997) computes its value to have a mean of 0.11 and a median of 0.03.

tations that much more difficult. However, we see no virtue in pretending that evolution is a simpler process than it actually is. Many evolutionary biologists fear that an overly simple conceptualization of the evolutionary process has, in some cases, led to erroneous conclusions being drawn (Coyne and Berry, 2000; Lloyd and Feldman, 2001). Yet modern evolutionary biology has much more to offer enthusiasts than the suggestion that the process of evolution is complicated. There are rigorous methods for detecting the action of natural selection (Endler, 1986a), for isolating characters (Wagner, 2001), for determining whether a character is an adaptation (Sinervo and Basolo, 1996; Orzack and Sober, 2001), and for drawing inferences about how characters have evolved (Harvey and Pagel, 1991) that could beneficially be used more frequently within evolutionary psychology. Rather than remaining content to rely on polemical assertions or deductive reasoning, evolutionary psychologists could directly evaluate their claim that there is little ongoing selection in modern human populations by utilizing well-established methods for estimating selection gradients and contributions to fitness (Lande and Arnold, 1983; Endler, 1986a). There is room for more evolution within evolutionary psychology.

Conclusions

It is clear that evolutionary psychology is a mixed bag. There are undoubtedly some very fine pieces of work that show genuine promise of being able to decipher the evolved structures of the human mind. The best of evolutionary psychology is as rigorous and sophisticated as any research carried out in the general area of human behaviour and evolution. However, the discipline is marred by a number of weak studies that do little more than use a Pleistocene stereotype to contrive a 'Just so' evolutionary story. Sadly, these poorer studies frequently have a sensational quality that results in their receiving considerable attention. Perhaps too much research in the field is a documentation of what is already known, accompanied by a *post hoc* evolutionary spin and a snappy press release. Other psychologists have stressed the need for more sophisticated theories than are typical of evolutionary psychology (e.g. Heyes, 2000).

It would be unfair to condemn the entire field of evolutionary psychology on the basis of the work of its weakest practitioners. The problems that are described in the previous sections are hardly irreparably damaging, and there is nothing to prevent evolutionary psychologists from using the EEA concept with greater caution, or paying greater attention to developments within evolutionary biology; indeed, some proponents clearly already do so. The evolutionary psychology perspective has brought the study of the mind well and truly into the domain of evolutionary theory, bringing with it a welcome focus on proximate mechanisms. It has proven an enormously creative approach to the study of human behaviour, and has introduced a wealth of new ideas and methods. Moreover, the evolutionary psychology literature has made important contributions to the understanding of culture (Sperber, 1996), decision making (Gigerenzer *et al.*, 1999; Todd, 2001), emotion (Fessler, 2002), language (Pinker, 1994), pregnancy (Profet, 1988; Fessler, 2002), psychological illness (Nesse and Williams, 1995), sexual behaviour and sex differences (Daly and Wilson, 1983; Miller, 1997), stigmatization (Kurzban and Leary, 2001), visual perception (Shepard, 1992), and many other topics (see Barkow *et al.*, 1992 or Barrett *et al.*,

2001 for comprehensive treatments). Yet for all the enthusiasm it has engendered, at this time evolutionary thinking makes up a very small component of psychological research. We believe that the likelihood of significant advances will be enhanced if evolutionary psychologists broaden their methodology to embrace other appropriate evolutionary perspectives, tools, and heuristics (Plotkin, 1994, 1997; Heyes and Huber, 2000).

There is one criticism of evolutionary psychology on which we have not yet dwelt, namely that it underestimates the critical role of cultural transmission processes in shaping human knowledge and behaviour. In the next two chapters we will consider evolutionary perspectives that treat culture as a much more dynamic and influential process than hitherto regarded. Maybe social scientists are right to view cultural processes as not always well specified by our genes or environment, and as having a limited autonomy from biological control. Perhaps culture is an important evolutionary player in its own right.

CHAPTER 6

Memetics

Daniel Dennett describes Darwin's theory of natural selection as like a universal acid that 'eats through just about every traditional concept, and leaves in its wake a revolutionised world-view' (1995, p. 63). As an explanatory abstraction, perhaps natural selection is simply too good an idea to be restricted to genes and biological evolution. As soon as *The Origin of Species* was published, scientists, philosophers, and social scientists inevitably began to speculate as to whether other entities, such as the central nervous system or scientific theories, might also be evolving by the same process. In *The Descent of Man*, Darwin himself described the evolution of language, stating boldly that 'the survival or preservation of certain favoured words in the struggle for existence is natural selection' (1871, p. 61).

Darwin's intuition that natural selection may be a general law for how a multitude of processes change has proven to be not unreasonable. The immune system generates antibodies through an equivalent selective process (Burnet, 1959), and there is a respectable scientific and philosophical tradition, somewhat esoterically known as evolutionary epistemology, which stresses the universal nature of natural selection (Plotkin, 1982; 1994), and is backed by luminary philosophers (Popper, 1979; Hull, 1982; Dennett, 1995) and Nobel Prize winning scientists (Lorenz, 1965; Edelman, 1987). Henry Plotkin's 1994 book, *Darwin Machines and the Nature of Knowledge*, is a compelling exposition of this perspective. Plotkin and others attempt to see which phenomena, in addition to selection on genes, can be fruitfully treated as selection processes.

While Darwinism continues to eat its way voraciously through countless academic disciplines, the social sciences stand out as a last hold of resistance. Stalwarts of the humanities have for years maintained that no biological theory is going to explain much about human cultural change. A great deal of what is interesting about humanity, it is claimed, cannot be explained in terms of genes or fitness. Evolution may help to explain what human beings have in common with other animals, but it is the differences that make us interesting and special. What can evolution possibly tell us about how human beings think or what they believe? Here, explanations in terms of culture are, for most social scientists, more compelling than biological accounts. What then if culture itself evolves?

In the final chapter of *The Selfish Gene* (1976), Richard Dawkins, explicitly dissatisfied with sociobiological explanations for human behaviour, let loose a new *cultural* replicator. Stressing the similarity between cultural and genetic transmission, Dawkins suggested that fashions, diets, customs, language, art, and technology evolve over historical time. He coined the terms 'replicator' and 'vehicle' to distinguish between the 'immortal' genes, which are replicated each generation, and the transient, vehicular organisms that house them. The gene is the archetypal replicator, but Dawkins proposed that a new, frequently insidious kind of replicator has recently emerged on this planet, a mind virus that infects us with catchy concepts and fashionable ideas.

> We need a name for the new replicator, a noun that
> conveys the idea of a unit of cultural transmission, or a

unit of *imitation*. 'Mimeme' comes from a suitable Greek root, but I want a monosyllable that sounds a bit like 'gene'. I hope my classicist friends will forgive me if I abbreviate mimeme to *meme*. (Italics in original; 1976, p. 206.)

Dawkins described how, just as soon as the genes had blessed this particular species of 'lumbering robots' with an enhanced capacity for imitation, the memes set in. There was, as Dennett puts it, 'an invasion of the body snatchers' (1995, p. 342). Memes have been described as having parasitized our vulnerable brains, turning them into vehicles for their own virulent propagation.

According to Dawkins (1976), memes possess variation, heredity, and differential fitness, the three characteristics that are necessary for evolution. They also commonly exhibit the qualities of particularly effective replicators – *longevity* (they frequently stay in our heads for long periods), *fecundity* (they can be copied and spread rapidly), and *copying fidelity* (at least some core components of some memes are reasonably faithfully reproduced). In the rich environment of human minds, these characteristics may be all that memes need to evolve. Dawkins suggested that meme evolution is not merely a process that can be metaphorically described in evolutionary terms, it *is* evolution by natural selection. In the benign environments of our conformist and indoctrinable minds, memes may compete against each other for our attention and acceptance. Only the fittest meme, for example the most memorable catchphrase, the trendiest fashion or the most comforting idea, will win the battle for our favour, and spread itself by tricking us into becoming its advertising agent. According to Dawkins, 'memes propagate themselves in the meme pool by leaping from brain to brain via a process which, in the broad sense, can be called imitation' (1976, p. 206). He suggested that we don't pick or choose our ideas and beliefs; on the contrary, they pick and choose us, and manipulate us to their own ends.

Dawkins described how an idea like a belief in a god can have a stability and penetrance in the cultural environment because of its great psychological appeal, as it 'provides a superficially plausible answer to deep and troubling questions about existence' (1976, p. 207). In another example, Dawkins suggested that a gene for celibacy may be very unlikely to spread among humans; however, a meme for celibacy could be successful if marriage and children detracted from a priest's capacity to influence his flock, so that celibate priests were more effective in passing on their ideas. Dawkins also proposed that co-adapted complexes of memes could evolve; that is, a collection of ideas such as an organized religion or a political party, which assist each other in mutual propagation. What was distinctive about the explanations that Dawkins was able to give for cultural phenomena was simply that 'a cultural trait may have evolved in the way that it has simply because it is *advantageous to itself*' (italics in original; 1976, p. 214). Dawkins suggested that we may not need to look for conventional biological survival values, or even functional explanations, for many human traits. Particular memes may thrive because they are good at spreading, that's all.

As scientific concepts go, the 'meme' meme had the best possible start—it was launched in one of the most popular scientific books of the twentieth century. Yet while computer geeks ran away with the idea, generating a popular subculture of meme followers, in academic circles the meme fell on fallow ground. With notable exceptions, such as the philosopher David Hull, anthropologist Bill Durham, and neuroscientist Juan Delius, the meme concept did not take off as an explanation for cultural phenomena. Why was this? The philosopher Daniel Dennett provided an answer that has more than an inkling of truth about it:

> I suggest that the meme's-eye view of what happened to the meme meme is quite obvious: 'humanist' minds have set up a particularly aggressive set of filters against memes coming from 'sociobiology', and once Dawkins was identified as a sociobiologist, this almost guaranteed rejection of whatever this interloper had to say about culture—not for good reasons, but just in a sort of immunological rejection. (1995, pp. 361–2)

In spite of his suspicion that Dawkins was retreating on the meme concept, and in the face of hostile criticism from the social sciences, Dennett set out in the 1990s to reinvigorate the meme. In 1991, *Consciousness Explained* was published, in which Dennett made memes the centre-piece of a grand theory of the evolution of mind:

> The way in which culture has become a repository and transmission medium for innovations ... is important for understanding the sources of design of human consciousness, for it is yet another medium of evolution. (1991, pp. 199–200)

Dennett argued, somewhat disturbingly, that the human mind is an artefact created by memes for memes. Four years later, Dennett produced *Darwin's Dangerous Idea* (1995), another bestseller, this time a more general advocation of universal Darwinism, and again with memes as a central concept. Dennett also stressed that memes possess variation, heredity, and differential fitness. For instance, not all tunes are the same (variation), some tunes we adopt and reproduce by singing or playing them (heredity) and tunes vary greatly in how memorable they are (differential selection). Dennett strongly defended the idea of cultural evolution, illustrating how culture changes over time, accu-mulating and losing features, while also maintaining features from earlier ages. Human language, written or spoken, was suggested to provide a suitable medium for meme transmission.

As minds are in limited supply, there may be considerable competition among memes for entry into as many minds as possible. Dennett describes a number of good tricks that he says memes have to disable opponent memes and win brain property rights. The ultimate trick is perhaps provided by the 'conspiracy theory' meme, which has a built-in response to the objection that there is no good evidence of the conspiracy: 'Of course not—that's how powerful the conspiracy is!' (Dennett, 1991, p. 206). Another possible example of how a meme may increase its circulation is the meme of including in a chain letter a warning about the terrible fates of those who have broken the chain in the past. A vivid illustration is provided by the 'St Jude' chain letter, which Oliver Goodenough and Richard Dawkins published in *Nature* in 1994, as a *bona fide* example of a mind virus. It begins:

With Love All Things are Possible

This paper has been sent to you for Luck. The original is in New England. It has been sent around the world. The Luck has been sent to you. You will receive good luck within 4 days of receiving this letter pending in turn you send it on. This is no joke. You will receive good luck in the mail. Send no money. Send copies to people you think need good luck. Do not send money cause faith has no price. Do not keep this letter. It must leave your hands within 96 hrs. An A.R.P. officer Joe Elliot received $40,000,000. George Welch lost his wife 5 days after this letter. He failed to circulate the letter.... St Jude.

Even Goodenough and Dawkins, hard-nosed evolutionary minded academics fully conversant with the corrupting
potential of memes, confessed to experiencing waves of
mild, irrational anxiety on deciding not to comply!

Dennett's writings reinvigorated memetics. Further
popular books followed, including Aaron Lynch's (1996)
Thought Contagion: How Belief Spreads through Society, and
Richard Brodie's (1996) *Virus of the Mind: The New Science
of the Meme*, which expanded on Dawkins's and Dennett's
ideas. In 1997, a new internet journal was created, called the
*Journal of Memetics: Evolutionary Models of Information
Transmission*, as a fresh forum for the publication of academic work on memes. In the late 1990s, the first academic
conferences on the topic of memetics were held, providing
further evidence that memetics was beginning to emerge as
an active research programme.

Susan Blackmore's (1999) *The Meme Machine* has been
the latest bestselling book to attempt to infest the planet
with this mind-warping virus. Blackmore's book was a
tour de force of memetic reasoning. Memes were suggested
to provide novel explanations for phenomena as diverse as
the origins of the large human brain, the emergence of
language, the existence of altruism, and the ubiquity of
New Age cults. More so than any book that preceded
it, Blackmore's meme-fest illustrated the potential of
memetics to tackle a range of issues central to science and
the humanities. According to Blackmore, not only could
memes account for cultural change, but they could also
drive genetic evolution through meme–gene coevolution.
Could memetics be, as Blackmore suspects, 'the grand new
unifying theory we need to understand human nature'?
(1999, p. 9). Or is it, as some critics suggest, just more
evolutionary story-telling?

Key concepts

In this section, we take a closer look at the questions 'what is a meme?', 'how do memes spread?', and 'are memes replicators?' We begin with a description of the meme's-eye view.

Taking the meme's-eye view

In *The Meme Machine*, Blackmore challenged us to '[i]magine a world full of hosts for memes (e.g. brains) and far more memes than can possibly find homes. Now ask, which memes are more likely to find a safe home and get passed on again?' (1999, p. 37). The answer given is the eye-catching or high-profile memes rather than the memes we might objectively judge to be in our interests, or worthy of our attention. From an evolutionary perspective, the only virtue is sheer replication. Blackmore was encouraging us to 'take the meme's-eye view' (1999, p. 37) to obtain an alternative perspective on the ideas that are stored in our minds. We are all used to thinking of our thoughts and beliefs as carefully selected or constructed by us, according to our particular disposition and powers of reasoning—but perhaps we've been taken in. Apparently, our minds are just a 'dungheap in which the larvae of other people's ideas renew themselves' (Dennett, 1995, p. 346).

With a direct correspondence to the 'gene's-eye view', the 'meme's-eye view' is the notion that, if we wish to understand what cultural phenomena ought to evolve, it is a convenient and useful heuristic to look at the problem from the perspective of the meme, and ask which properties would be most likely to increase its frequency. Dawkins suggested that in the same way that blind natural selection makes genes behave as if they were 'active agents, working purposefully for their own survival, perhaps it might be convenient to think of memes in the same way' (1976, p. 211). Dennett summed up this perspective with typical panache:

> A scholar is just a library's way of making another library. (1991, p. 202)

This is the perspective that distinguishes memetics from alternative approaches to understanding culture, and from some earlier research traditions within the social sciences, known as 'information transfer' and 'diffusionist' schools. Meme enthusiasts believe that cultural traits evolve, not because they are of utility to individuals (although they may be), but because they aid meme propagation. They are there for the good of the memes. There is no necessary relationship between a meme's replicative capacity and its contribution to our fitness, although Dennett suggests that 'What memes provide in return to the organisms in which they reside is an incalculable store of advantages—with some Trojan horses thrown in for good measure, no doubt' (Dennett, 1995, p. 365). Some memes (like dancing) may promote health and happiness. In contrast, others (like warmongering) may reduce our chances of survival. Dawkins (1976) noted that some of his colleagues wanted to drag memetics back to biological advantage, suggesting that some memes have great psychological appeal because our brains have evolved to choose them or their like. They wanted to find a way in which having a brain that chose particular memes improved gene propagation. However, Dawkins resisted any suggestion that memetic evolution need be subservient to genetic evolution. Once our brains evolved a capacity to copy others, memes were away and nothing could stop them flourishing. Blackmore's view is that a lot of memes may actually thrive precisely because they *do* contribute to genetic fitness, but that contributing to fitness is only one of many ways in which a meme can be replicated (Blackmore, personal communication).

Dennett (1991) suggested that when we study a trait, say music, we should begin by asking 'Cui bono', that is 'who benefits from this'? The conventional explanation would be that traits like music benefit us, enhancing our well being and state of mind. The sociobiological perspective might be that these traits are there for the benefit of genes; that is, genes that enhanced the capacity of our ancestors to make or appreciate fine music increased their frequency in the gene pool. The perspective from memetics provides a third alternative: music may be there solely for the benefit of music memes.

What is a meme?

In spite of the intuitive appeal of the meme, it has proven a very tricky concept to pin down. Exactly what is a meme? To describe it as a unit of culture is unsatisfactorily vague. In any case, there are problems with devising an adequate definition of culture. Even if everyone agrees that memes are replicators, what precisely is being replicated? Information? Instructions? Behaviour? Artefacts? Several commentators (e.g. Delius, 1991; Aunger, 2002) have argued that memes are complex neural structures, parallel to the identification of genes with complex structures of DNA, while Gatherer (1998) argues that memes are behaviour patterns and artefacts. However, both Dawkins and Dennett describe memes as ideas or information.

A description of memes as information would appear to be a popular, if not a consensus position. For Dennett, memes are instructions for behaviour embedded in human brains, and expressed in meme 'vehicles':

> [Memes are] carried by meme vehicles—pictures, books, sayings. . . . Tools and buildings and other inventions are

also meme vehicles. A wagon with spoked wheels carries
not only grain or freight from place to place; it carries the
brilliant idea of a wagon with spoked wheels from mind to
mind. (1991, p. 204)

Other commentators, for example, Blackmore (1999)
and Hull (2000) would include artefacts, such as books and
floppy discs, as extraneural stores of memetic information.

Much debate and confusion has centred around what, in
meme terms, is analogous to the genotype–phenotype
distinction between the genetic constitution of an organism (its genotype) and the characteristics of the organism
itself (its phenotype). Blackmore illustrates the complexities with the example of a meme for making pumpkin soup,
which can spread through spoken word, written recipe,
imitation of another soup-maker, or perhaps even reconstructing the recipe following soup consumption:

> What if I sent you the recipe in the post and you passed it
> on to your granny and she made a photocopy for her
> friend?… the instructions on making the soup might go
> from brain to piece of paper, to behaviour, to another
> brain, to a computer disk … Which is the genotype and
> which the phenotype in each case? Are we to count memes
> as only the instructions in the brains or the ones on paper
> too? Are the behaviours memes or meme-phenotypes? If
> the behaviour is the phenotype, what then is the soup?
> There are lots of possibilities in memetic evolution because
> memes are not confined by the rigid structure of DNA.
> (1999, pp. 61–2)

There would seem to be two internally consistent solutions to this impasse. Gabora (1997) discusses the genotype–phenotype distinction in terms of information and its
implementation. Gabora limits memes to mental representations and treats their implementation in behaviour or
artefacts as the phenotypes of these mental representations.

Perhaps a clearer distinction would be to describe behaviour as the phenotype (implementation in the vehicle) and arte-facts as the extended phenotypes (implementation outside the vehicle). Dawkins (1982) introduced the idea of the 'extended phenotype' to capture the notion that genes can express themselves outside of the bodies of the organisms that carry them. For instance, a beaver's dam is an extended phenotypic effect of the beaver's genes. The knowledge of how to make soup is the meme and therefore the gene ana-logue, while the soup-making behaviour is the phenotype analogue, and the soup the extended phenotype analogue. This is satisfactory provided it is recognized that some implementations, for example, a spoken recipe for making soup (phenotype) or written recipe (extended phenotype), are characterized by an explicit attempt to provide a syntac-tic account of the meme's informational content (i.e. the knowledge in the original cook's head), while other imple-mentations, such as the soup, or the cooking behaviour, are not. Blackmore (1999) distinguishes between *copy-the-product* and *copy-the-instructions* forms of meme transmis-sion. Learning to make soup by reproducing the behaviour of the cook or reverse-engineering the recipe from the soup would be examples of copy-the-product, while following a written recipe would be an example of copy-the-instructions. We suggest that there are in fact three logically distinct means of meme acquisition, copy-the-process (that is, reproduce the behaviour), copy-the-product (reverse-engineer the informational content of the meme), and copy-the-instructions (exploit a syntactic depiction of the meme).

The second approach, advocated by Hull (2000), is to describe as replication all cases in which information is passed along largely unchanged regardless of whether the substrate is a brain (i.e. the cook's idea) or artefactual (e.g.

the written recipe). Replicators are distinguished from interactors (loosely synonymous with Dawkins' 'vehicles'),[1] which are the entities that exhibit adaptations (e.g. the soup), but are characterized by a loss of information. In this example, the loss of information is illustrated by how difficult it is to reconstruct the recipe from the soup alone. Here, a good photocopy of the soup recipe would be a replicator and gene analogue, but a bowl of soup would be an interactor and phenotype analogue. This is satisfactory provided we allow entities to be simultaneously replicators and interactors (as, indeed, genes are), and for replicators to exhibit adaptations (such as the written recipe). Whatever solution is adopted, memeticists will have to accept that replicators and vehicles will operate differently at genetic and cultural levels.

How do memes spread?

In *The Selfish Gene*, Dawkins paid little attention to the psychological processes that underpin meme transmission. He merely stated, '[i]mitation, in the broad sense, is how memes *can* replicate' (italics in original, p. 208). The qualification *in the broad sense* is perhaps designed to distinguish the processes that Dawkins had in mind from the more narrow use of the term 'imitation' employed by social

---

[1] Organisms are commonly regarded as: (1) phenotypes that interact with their environments, that survive, reproduce, and pass on genes and (2) the entities that are 'produced' by genes. Hull (personal communication) argues that his notion of 'interactor' is significantly different from Dawkins' 'vehicle'. As in (1), Hull's regards his concept as a populational notion (the population of entities that directly interact with the external environment), while Dawkins's 'vehicle', as in (2), is regarded as more embryological (vehicles are the entity that replicators produce).

learning theorists. The colloquial or broad use of imitation refers to any process by which individuals learn from others —*social learning* would be the phrase used by psychologists. The narrower, academic usage of imitation pertains exclusively to those cases in which an individual learns to perform a motor pattern in a particular context, or for a particular consequence, through observing another individual do the same. When we copy someone making soup, or follow written or verbal instructions, or hum the songs they played on their guitar, or admire the clothes they wear, we are not attending to and reproducing their precise bodily movements (narrow imitation), we are engaging in an unspecified form of social learning.

Social learning theorists distinguish between a multitude of processes that can result in one individual learning something from another, with obscure labels such as *local enhancement, goal emulation, matched-dependent learning,* or *opportunity teaching*; imitation in the narrow sense is just another one of these. All of these processes are capable of resulting in the transmission of novel-learned traits through a population. Which of these, then, can underlie meme transmission? Some researchers (e.g. Blackmore, 1999) argue that meme propagation is mediated exclusively by imitation (in the narrow sense), while others (e.g. Reader and Laland, 1999) maintain that any process that results in the stable transmission of learned information could support memes. One repercussion of this debate is that non-human animals learn from each other frequently, but rarely through imitation. If Blackmore is correct, then humans may be one of just a handful of species that have memes.

So do other animals have memes? Dawkins (1976) began his chapter on memes with the example of a bird, the New

Zealand saddleback, that has song dialects that vary from one locality to the next, where each young male learns the song of its neighbours, and where song patterns change in frequency over time. Dawkins states that 'song in the saddleback truly evolves by non-genetic means' (1976, p. 204). Is bird song learning imitation in the narrow sense? This is a bone of contention, even for social learning researchers, some of whom distinguish between vocal and motor imitation, claiming that the former is easier. There is a sense in which the question is irrelevant: we know that bird song exhibits variation, heredity, and differential transmission between individuals, so why not describe this evolution as memetic?

In fact, some of the neatest empirical work on memes has been that which applied the new replicator concept to bird song (Alejandro Lynch, 1996). Variation in songs both within and between populations has been studied using the idea that bird song syllables, phrases or entire songs can be regarded as cultural units (Catchpole and Slater, 1995), and mathematical models of population genetics can be co-opted to study meme evolution (Alejandro Lynch, 1996). Much of this work has been carried out on chaffinch songs. Studies have revealed that separate populations of chaffinches differ slightly in the composition of their songs, with island populations having particularly high numbers of alternative memes. Young male chaffinches bring songs with them when they migrate from natal areas, and also copy the songs of resident males, although with some copying errors. Studies of a single chaffinch population have found that most of the song variants within the population had changed over a period of around 20 years, probably as a result of immigration and high rates of copying error. These copying errors can be thought of as mutations that result in new memes. Some of the new song types failed to be copied and became extinct. In chaffinches, which song memes were copied and which maintained appeared to be random. Thus, cultural evolution of bird songs appears to be ongoing, although selection favouring memes with particular qualities has not been demonstrated. The memetic divergence of populations appears to occur through chance processes, analogous to random genetic drift. An exciting aspect of this area is the possibility of relocating birds and their memes to new geographical areas to study the spread and acquisition of memes or meme complexes. The memetic approach has clearly already increased our understanding of variation in bird songs over time and space, by viewing this as the product of cultural evolution.

If memes can be made to do good work on bird song, and perhaps on the protocultural traditions of other non-human animals, then this is a powerful argument for adopting a liberal stance on memes. Debating whether we should restrict memes to humans and to imitation, Hull (2000) writes:

> For now, I would think that casting our net too broadly is a better strategy than casting it too narrowly. (p. 2)

Ultimately, it is an empirical issue as to which processes support meme transmission. This is one area in which meme enthusiasts can make a genuine contribution through experimentation on social learning processes. If Dawkins is correct, and the qualities that make for high survival value among memes are longevity, fecundity, and copying fidelity, then this insight will be of value to researchers interested in the stability of human and animal traditions.

Meme fidelity

A major question mark against memes, to which both Dawkins and Dennett allude, is whether they have sufficiently high copying fidelity, or accuracy of reproduction. When discussing meme fidelity Dawkins confesses 'here I must admit that I am on shaky ground' (1976, p. 209), and he acknowledges, as an example, that his ideas published in *The Selfish Gene* resulted from a blending of Trivers's and his own memes. Similarly, Dennett (1995) asks: 'Isn't one of the hallmarks of cultural evolution and transmission the extraordinarily high rate of mutation and recombination?' (p. 355). If memes are constantly passed on in altered forms, can they be described as replicators? This looks quite unlike the particulate, virtually error-free copying of gene translation. At first sight, meme evolution appears so fluid, subject as it is to continuous mutation, blending of memes, and cross-fertilization between lineages, that it is difficult to see how it could generate complex adaptations analogous to the vertebrate eye or hand.

There are at least two counterarguments that have been put forward. The first was expressed most clearly by Dawkins:

> It is possible that this appearance of non-particulateness is illusory, and that the analogy with genes does not break down. After all, if we look at the inheritance of many genetic characters such as human height or skin colouring, it does not look like the work of indivisible and unblendable genes. (1976, p. 209)

Conceivably, it may be fruitful to regard complex cultural structures, or memeplexes, as composed of subelements of particulate, interchangeable memes. For example, the notion of space travel is a complex of ideas (pertaining to rockets, astronauts, moon landings, extra-terrestrials, etc.), which vary from individual to individual depending on which discrete elements they have. What appears to be blended are not the replicators but the vehicles, when they are the expression of multiple mix-and-match memes. If this is the case, memeticists need to develop and employ blending models of meme transmission, analogous to the theoretical models used by animal breeders and in quantitative genetics. In fact, a form of quantitative memetics has already been developed by researchers engaged in cultural evolution and gene–culture coevolution: this is described in the following chapter.

The second counterargument is that, while every version of a meme varies from one person to the next according to each individual's personal experiences, all memes have a core element that is shared knowledge. For instance, the core element of the *wheel* meme might include the knowledge that it is round and that it rotates about a central axis, whereas whether or not it happens to be blue, made of wood, or have spokes is not central. There are two important caveats to this argument. First, individuals can only be said to share the same meme if they acquired its core element directly or indirectly from each other or from a common third party (Hull, 2000). In genetic parlance, to be counted as the same, memes must be *identical by descent.* Secondly, we should expect the core element of a meme to change over time, since cultural evolution is a dynamic process. Today the core element of the *evolution by natural selection* meme is very different from that of Darwin's day, including as it does knowledge of Mendelian genetics and DNA.

Sperber (1996, 2000) has criticized the notion of meme transmission, arguing that cultural information is not

transmitted from individual to individual but is reconstructed by each individual inside their heads, perhaps
exploiting evolved psychological mechanisms. The fact that
the same memes appear to crop up in different individuals
following social interaction may be a manifestation of the
fact that individuals have independently reconstructed the
same idea in their heads, both because they have similar
brain structures (with evolved predispositions to adopt
some memes over others), and because only certain types of
idea are stable over extended periods of time (for example,
easily remembered ones). Plotkin (2000) argues that some
'deep memes', acquired over long periods of enculturation,
have this stable quality. Sperber's argument is only a problem for memeticists if they decide that reconstructed
memes are not memes. As all memes are likely to have at
least a degree of reconstruction (Sperber, 2000; Bloch,
2000), such a stance would appear foolhardy.

Indeed, memeticists seemingly have grounds to predict
the existence of reconstruction. To the extent that meme
messages contain information that is likely to be accurately
reconstructed by receiving brains anyway, then they contain
redundancy. However, natural selection eliminates redundancy, as organisms that do not waste energy and resources
will generally outcompete those that do. Thus we might
anticipate that memes only transmit that which is absolutely necessary, and leave the rest for the receiving mind to
reconstruct. Provided reconstruction is socially contingent,
that is, provided that the expression of a meme by one
individual provides the trigger for a second individual to
acquire the same meme as Sperber suggests, then reconstructed memes are just as subject to cultural evolution as
transmitted memes. If Sperber's criticism is correct it is
actually good news for memetics, since if we are predis-

posed to reconstruct particular memes it will mean there is another reason to expect meme fidelity to be high. The important point is that the fact that memes may be reconstructed, perhaps according to the directive evolved genetic predispositions, does not detract from the hypothesis that culture evolves. The degree to which information is transmitted or constructed during social learning is wide open for study and, again, it is an area of research that memeticists can contribute to through experimentation.

Case studies

In this section, we illustrate how thinking from the perspective of the meme has provided novel views on religion, consciousness, and changes in scientific theories.

Religion

One of the most controversial applications of memetic reasoning has been to account for religion. An organized and socially sanctioned belief in a god is to many people a given and a truth. This belief is not always regarded as something that is a legitimate focus for scientific enquiry. Even among non-believers, the idea that religions could be self-serving and self-perpetuating ideational complexes that hoodwink us into spreading their message is somewhat disturbing. Yet that is precisely what they have been argued to be by advocates of the meme's-eye view.

This infamous account was first proposed by Dawkins in *The Selfish Gene* (1976), and elaborated in later writings. Dawkins argued that cultural selection would favour memes that gang up effectively into super-attractive *co-adapted meme-complexes*, or *memeplexes* (Speel, 1995; referenced in Blackmore, 1999). Dawkins suggested that we could regard a church, with its architecture, rituals, laws, music, art, and written tradition, as just such a memeplex. He argued that the idea of a god and the religion memes that aggregate around it replicate themselves by providing convincing answers to life's great questions.

Religions, however, are perhaps much more sinister than that. Dawkins suggested that they appear to employ various tricks, and co-opt other memes that facilitate their replication by the most dastardly of connivances. For instance, according to Dawkins:

> an aspect of doctrine which has been very effective in enforcing religious observance is the threat of hell fire. Many children and even some adults believe that they will suffer ghastly torments after death if they do not obey the priestly rules. This is a particularly nasty technique of persuasion, causing great psychological anguish ... The idea of hell fire is ... *self-perpetuating,* because of its own deep psychological impact. It has become linked with the god meme because the two reinforce each other, and assist each other's survival in the meme pool. (Italics in original; 1976, p. 212)

Then there is faith:

> [Faith] means blind trust, in the absence of evidence, even in the teeth of evidence ... The meme for blind faith secures its own perpetuation by the simple unconscious expedient of discouraging rational enquiry. (Dawkins, 1976, pp. 212–13)

In fact, consider every possible trick that memes could employ to increase their frequency and memeticists suggest that such tricks are observed among organized religions (Aaron Lynch, 1996; Blackmore, 1999). They point out that memes would thrive that encouraged credit and praise to be heaped on individuals who read or learn verbatim texts describing the meme-complex; for example, the learning of Bible stories. Children adopt their parents' memes, hence specific religious memes may encourage having children, discourage abortion or contraception, encourage respect for elders, and discourage marriages between faiths. Memes could increase their frequency through conversions, so the most effective religions would be expected to place a premium on evangelicalism, proselytism, missionary work, and punishment of non-believers. Additionally, any challenge to the meme-complex might be treated extremely severely as, for example, in the case of Ayatollah Khomeini's fatwa on the author Salman Rushdie.

Blackmore (1999) asks her readers to reflect on why some minor religions went on to become great faiths, while the majority died out with the death of their leader. Her answer is that, of the many religious ideas, only some had packages of memes that were effective gimmicks for propagation, with particularly compelling (and difficult to disprove) explanations for life, and these became the major religions. Citing the work of theologian Hugh Pyper, Blackmore describes the Bible as the fittest of all books. She writes:

> Western culture is the Bible's way of making more Bibles. And why is it [the bible] so successful? Because it alters its environment in a way that increases the chances of it being copied. It does this, for example, by including within itself many instructions to pass it on, and by describing itself as indispensable to the people who read it. It is extremely adaptable, and since much of its content is self-contradictory it can be used to justify more or less any action or moral stance. (1999, p. 192)

Attributing motives to memes is simply an intellectual stance adopted to help envisage which memes might be expected to have evolved. As Blackmore explains, religious memes did not, indeed could not, set out to succeed. She suggests that they were simply ideas and behaviour that had some utility in explaining the world and succeeded where others failed because they had the right combination of mutually supportive ideas that allowed them to be repeatedly passed on. It is worthy of note that there are other evolutionary approaches to understanding religion, many of which stress the advantages that religion bring to the individual (e.g. Hinde, 1999). Later in the chapter, we will see that one criticism of memetics is that it underplays the selective role of brains in choosing which memes are adopted.

Consciousness

At the end of *The Selfish Gene*, having detailed the artful, manipulating nature of both genes and memes, Dawkins leaves us with one crumb of comfort:

> We are built as gene machines and cultured as meme machines, but we have the power to turn against our creators. We, alone on earth, can rebel against the tyranny of the selfish replicators. (1976, p. 215)

A few years later Dennett was to take even that away. For Dennett, the human mind is itself an artefact created when memes restructure the human brain to make it a better habitat for memes:

> If it is true that human minds are themselves to a very great degree the creations of memes, then we cannot sustain the polarity of vision with which we started; it cannot be 'memes versus us,' because earlier infestations of memes have already played a major role in determining who or what we are. The 'independent' mind struggling to protect itself from alien and dangerous memes is a myth. (1991, p. 207)

Dennett provided an alternative means of interpreting consciousness by taking the meme's-eye view: he suggested that our decision making processes are guided by our memes. Consciousness is frequently thought of as something uniquely human and special, but whether it provides any selective advantage has been fiercely debated. Some argue that consciousness could not have evolved unless it served a function to the individual. We might envisage that a sense of 'self' might benefit animals by ensuring they take better care of their bodies or perhaps by making it easier to predict the behaviour of other individuals, or even to deceive or outwit others. In comparison, some researchers suggest that consciousness has arisen as a by-product of selection for some other capability, such as intelligence. However, Dennett argues that human 'consciousness is *itself* a huge collection of memes (or more exactly, meme-effects in brains)' (1991, p. 210). While he concedes that there are possible benefits to us of having consciousness, he points out that we cannot entirely rule out the possibility that consciousness plays no essential role in how our brains work. At least 'some features of consciousness may just be selfish memes' (1991, p. 221). He suggests that consciousness may be nothing more than an inner self-representation, something that requires no special explanation, and which is produced by memes for the benefit of passing on those memes. As such, a robot could be provided with memes that allow it a conscious idea of itself.

In contrast, Blackmore (1999) concentrates on the idea of one's inner 'self' and suggests that the 'self' is merely a collection of memes that has accumulated over a lifetime. Human subjectivity (what is it like to be me now) is structured by our memes, while other forms of consciousness can be attained without reference to this 'self'. Memes make use of the 'self' by calling themselves 'beliefs', 'likes', or 'dislikes'. Those memes that make their vehicles argue for them, fight for them, or press them upon others, will be more successful at spreading themselves. Blackmore suggests that:

> The self is a great protector of memes, and the more complex the memetic society in which a person lives, the more memes there are fighting to get inside the protection of the self. (1999, p. 233)

Memes that are associated with a person's self-concept can gain an advantage in terms of being replicated. Blackmore suggests that memes 'force genes to build ever better and better meme-spreading devices' (1999, p. 119), such as big brains, language, and intelligence, and that meme 'competition drives the evolution of the mind' (1999, p. 17). This meme competition leads to us acquiring more knowledge, opinions, and beliefs of our own that strive to convince us that there is a real self at the centre of it all.

Science

The great 20th-century philosopher, Karl Popper, was among the first to suggest that the scientific process operated on Darwinian principles. A variety of scientific theories and hypotheses are proposed by scientists, and those theories that are confirmed through experimentation and found most useful by other scientists triumph in the battle to monopolize academic minds. Another philosopher of science, David Hull, has developed these ideas further in a series of perceptive essays (e.g. Hull, 1982; 1988; 2000).

Hull (1982) not only distinguishes between *replicators*, which are the entities that pass on their structure largely intact, and *interactors*, which are those entities that exhibit adaptations, biasing replication because of their relative success in coping with their environments, but also defines *lineages*, which are evolving collections of interactors that share replicators. In genetic evolution, genes are the replicators, organisms are the interactors, and species are the lineages. In conceptual evolution, memes are the replicators, human beings are the primary interactors with their environments, and conceptual systems are the evolving lineages.

Hull believes that scientific communities (e.g. Darwinians) are a collection of interacting scientists that have in common one or more memes (e.g. natural selection, Mendelian genetics, etc.) that are expressed in an evolving conceptual system (e.g. Darwinism). Researchers of today that are part of the Darwinian community have different views from their 19th-century counterparts. What unites them is the notion that they derived their beliefs from preceding Darwinians. But how can we tell whether a scientist is part of a scientific community? According to Hull (1982), in exactly the same way we can tell whether an individual organism is a member of a particular species:

> Pick an organism, any organism, and invent a name. Any
> other organism with the appropriate relations to this type
> specimen belong in the same species with it. The
> appropriate relations are gene transmission and gene
> exchange, not similarity. The type specimen is in no sense
> 'typical'. It is merely one node in the geneological nexus.
> Scientific communities can be individuated in the same
> way. Pick a scientist, any scientist, and trace out his
> scientific connections. If he belongs to a community,
> the contours of that community will materialize. Any
> scientist who actually belongs in a community can serve as
> the 'type specimen' ... of that community. He need not be
> the most important member of that community, nor even

the scientist after whom the community is named.
(p. 297)

Hull suggests that, to belong within the same lineage, scientists must have gained their information from each other, rather than merely holding similar views. Once such communities of scientists are defined, an evolutionary analysis of the development of ideas can begin. In fact, Hull argues that science is analogous to artificial selection rather than natural selection:

> Just as the breeder consciously selects the organisms that he breeds in order to produce desired changes in his stock, the scientist chooses conceptual variants in order to improve his scientific theories. Both processes involve conscious, intentional choices even though many of the results in both cases may be unanticipated. (1982, p. 317)

Hull is also one of the few people to have carried out empirical work on memes (Hull, 2000). For instance, if particular scientific communities can be regarded as equivalent to species, then Hull reasoned we ought to see a difference between how scientists respond to the ideas of members of the same and different communities, analogous to the competition that is seen within species and between species. Hull studied two research groups of biologists, both of which work on the classification of organisms: 'numerical taxonomists' and 'cladists'. These research groups differ in the methods that they use: numerical taxonomists attempt to classify organisms by measuring how similar or different they are in all of their physical characteristics, while the cladists look for whether organisms share specialized or novel characteristics in addition to more common ancestral features. Hull wanted to know whether members of these communities treated fellow members of their group differently from scientists working in the other research group.

To test his hypothesis, he studied the referees' reports on all of the papers submitted to *Systematic Zoology*, the leading journal in the field, over a 5-year period. He examined whether the cladists treated papers submitted by numerical taxonomists differently from those of fellow cladists; for example, whether one or other group's papers were rejected more frequently. Initially he found no such effect, but Hull subsequently realized that at the time that he had conducted his study a 'speciation event' was occurring, with cladists splitting into 'transformed cladists' and 'phylogenetic cladists'. When he went back to his data and reanalysed it as three species, Hull found the pattern that he had predicted, with scientists being harder on papers submitted by members of a research community other than their own.

Critical evaluation

Most of the evolutionary approaches that we describe in this book have been subject to considerable criticism. A lot of the time the critics have had something of a case, with the overenthusiastic evolutionary minded often being guilty of speculative and amateur story-telling. Even the most avid meme devotee would have to admit that meme advocates have produced more than their fair share of unsupported narratives. However, much of the criticism of memetics has concentrated on apparent 'problems' that on closer examination seem to disappear. In this section we first evaluate these commonly sighted problems with memetics, and then go on to consider two criticisms of memetics that are less easy to dismiss. We will see that there has been little emphasis on the possibility that humans may have evolved predispositions to prefer some memes to others. Moreover, while memetics has provided a new way of looking at the world, there is little solid evidence with which to evaluate the usefulness of the meme concept. Enthusiasts have carried out precious few experiments or empirical studies, and even some of its leading advocates question whether it can ever become a science (Dennett, 1995). We will end by providing some ideas about how culture evolution could be studied in the future.

Commonly cited 'problems' with memetics

Virtually all of the commonly cited 'problems' of memetics are either irrelevant or reservations that apply equally to biological evolution. Hull (1982) is surely justified in asserting:

> One should not expect more of a theory of sociocultural evolution than one does of a theory of biological evolution. (p. 277)

Let us consider three of these allegations in the next sections.

1. You can't define the boundaries of a meme

If genes are regarded as clean, particulate pairs of alleles that reside at an easily definable locus on a well-charted chromosome and species are conceived as self-apparent natural kinds, then in contrast meme boundaries will appear disturbingly fuzzy. However, memetic units only appear hazy in comparison with these simplistic concepts of gene and species. In fact, the biological world is a lot messier than commonly conceived.

In spite of extraordinary progress in mapping the human genome, biologists are still unable to say with complete certainty where on a chromosome a gene starts, where it ends, and which sections in between should be regarded as the gene. Mendelian genes are not all that particulate and numerous alternatives to Mendelian diploid inheritance exist. Molecular biologists have uncovered regulatory genes, introns, exons, junk DNA, mitochondrial DNA, chloroplast DNA, selfish DNA, transposable elements, retroviruses, and countless other complications. Biological species appear self-evident but over the years they have also proved tricky to define. According to Mayr's (1942) *biological species concept*, species are not defined as belonging to a particular type by virtue of their sharing particular physical characteristics, but are defined as a group of populations that actually or potentially interbreed. However, the boundaries between species then become less clear cut: not all species are sexual, not all sexual species have two sexes, not all matings are within species, not all 'inter-specific' hybrids are sterile (see Hull, 2000 for an extended discussion of this issue).

If uncertainty about the boundaries of a gene or species has not prevented progress in evolutionary biology, why should the fuzzy units of memetics be regarded as problematic? Dawkins provided a simple and operational means of delineating memes:

> [If a meme] can be subdivided into components, such that some people believe component A but not component B, while others believe B but not A, then A and B should be regarded as separate memes. If almost everybody who believes in A also believes in B—if the memes are closely 'linked' to use the genetic term—then it is convenient to lump them together as one meme. (1976, p. 210)

Similarly, Hull (1982) has provided an operational method for delineating conceptual lineages. We suggest that memeticists should just get on with it.

2. Memes merge together

> Biological evolution is a system of constant divergence
> without subsequent joining of branches. Lineages, once
> distinct, are separate forever. In human history,
> transmission across lineages is, perhaps, the major source
> of cultural change. (Gould, 1991, p. 65)

Gould's point, that memes leap across conceptual lineages, was regarded as particularly troubling by Dennett for two reasons: first, because lines of descent become hopelessly muddled; and, secondly, because the outward expression of memes changes so fast that there is no chance of keeping track of particular memes. These concerns were sufficiently onerous for Dennett to express pessimism over whether a science of memetics could become established.

In fact, this is another instance where an overly simple model of biological evolution is employed as a standard against which to dismiss a 'complicated' memetics. Biological lineages have repeatedly come together over evolutionary time. While it is rare for two distinct biological species to merge into a single species, a process known as introgression, this does in fact occur naturally. Much more common is the horizontal transfer of genetic material between species as a result of the action of viruses and plasmids. More frequent still are symbiotic associations between species. What are lichens if not the coming together of previously distinct algae (or cyanobacteria) and fungi into a merged identity? What are obligate mutualisms if not two separate species whose destiny has become intertwined? Many of the major transitions in evolution, such as the origin of eukaryote cells or multicellularity, are thought to have resulted from the ganging up of lower level entities into a higher level. The fact that gene-complexes come repeatedly together and split up again over evolutionary time scales has not prevented population geneticists and molecular evolutionists from studying genes. Hence, why should similar complications paralyse research into memetics? Tracking memes down conceptual lineages will not be easy, but Hull (1982; 2000) has not only provided a methodology for doing this, but has actually done it.

3. Memetic evolution is Lamarckian and directed

It is frequently asserted that cultural evolution is Lamarckian, and sometimes this is in itself regarded as sufficient to invalidate it (Hull, 1982). The term 'Lamarckian' is employed to depict instances where acquired characteristics are inherited, and this process has long been discredited as playing any role in biological evolution. Clearly, memetic evolution is not literally Lamarckian since nothing that we learn results in a change in our genes. However, social learning can appear analogous to Lamarckian inheritance because individuals frequently inherit information that others have previously modified. Whether or not memetic inheritance is appropriately described as Lamarckian depends on how meme genotype- and phenotype-analogues are defined (see Blackmore, 1999, for a useful discussion). Some definitions of memes allow for an analogue of Lamarckian inheritance and some do not. The important point, however, is that, since no-one thinks that social learning is literally Lamarckian, this label cannot cast a slur over cultural evolution. Individuals certainly modify the information they acquire and whether one regards that as resulting from a Lamarckian process, a memetic mutation, or some alternative action, heritable cultural variants will still be subject to differential transmission and adoption. We suggest that digressions over Lamarckian inheritance are unnecessary.

We will also not be distracted here with discussions over the possible analogies between genetic evolution and memetic evolution as this question is addressed in the following chapter.

A related, but more interesting, point is that memetic evolution is sometimes directed and intentional. Hull (1982) notes that the characteristic that commentators have in mind when they claim that sociocultural evolution, especially conceptual development in science, is 'Lamarckian' is that at least sometimes people actually notice problems and try to solve them. For instance, Pinker states:

> Memes such as the theory of relativity are not the cumulative product of millions of *random* (undirected) mutations of some original idea, but each brain in the chain of production added huge dollops of value to the product in a nonrandom way. (Italics in original; cited in Dennett, 1995, p. 355.)

This is correct, but it does not invalidate cultural evolution any more than the direction and intention imposed by population geneticists that selected lines of *Drosophila*, or the animal breeders that favoured high milk yields in cows, invalidates describing the results of artificial selection as evolutionary. The fact that mutations at the cultural level are sometimes not random but *smart variants*, informed by information acquired through evolutionary and non-evolutionary processes at other levels, is not a weakness of memetics, but instead makes it a subject ripe for investigation.

How minds select memes

One sinister aspect of the meme's-eye view is that human beings seem to have been stripped of their ability to chose their own beliefs, values, and ways of life. Apparently, nefarious mind viruses are running our lives. The memes are choosing and manipulating us, not the other way round. Surely this surreal alternative perspective can't be the whole story? After all, our minds have evolved over millions of years. Wouldn't evolution at least have fashioned us with an ability to evaluate the alternative options and filter the available information that is adopted? If our bodies have an immune system to quell biological viruses, then shouldn't we expect our minds to have analogous defences to suppress rogue memes? The stance advocated by some memeticists may be missing some of the underlying complexity to human behaviour.

Aunger (2000b) identifies a key issue for memeticists to investigate: namely, whether the design in cultural 'adaptations' is best described as artificially selected by people to reflect their needs or as the unintended outcome of independent replicators. For instance, has the human brain been shaped to have certain properties that 'god' happens to fit, as suggested by Hinde (1999), or is the god concept merely a clever replicator, as Dawkins (1976) says? It is an empirical issue to establish whether we can best understand the concepts that we have in our heads as reflecting the past natural selection of genes expressed in brain development or the infectiousness of memes. We suspect that the reality lies somewhere between these poles.

Minds are not like vacant apartments to let, idly awaiting a meme to take up residence. The ideas, knowledge, and skills that we acquire from others are likely to reflect to a large extent the predispositions, capabilities, and beliefs that we have already. Certain individuals may be more likely to adopt one meme rather than another because of their genetic background. The recent work of behavioural geneticists suggests that genetic differences explain some of the variability in human personality traits and behaviour

(Eaves *et al.*, 1989; Bouchard *et al.*, 1990). We will see in the next chapter that this work is not uncontroversial, and can be criticized on a number of grounds. Nonetheless, the claim that there are some heritable genetic differences among people that influence behaviour is highly plausible. It is a widely held view within evolutionary biology that most variable traits show substantial underlying genetic variation. Plant and animal breeders are usually successful when they conduct a programme of artificial selection to increase milk yield in cows or some economically important trait of domestic plants, and this would not be possible without genetic variation in the selected character. These results hold for behavioural traits in animals, such as the tendency of rodents to explore a cage or pigeons to return home. Given that the propensity of people to adopt a particular meme could be affected by many aspects of brain chemistry and organization, and given that it is likely that such aspects of the brain are affected by many different genes, it is certainly credible that some of the variation in people's beliefs and behaviour is affected by genetic variation. Intuitively, it seems to make sense that individuals with a more intellectual bent will be more likely than others to take up chess, while sensation seekers will be predisposed to acquire the memes for scuba diving or hang-gliding. Moreover, if evolutionary psychologists are correct, then there will be many memes that are found universally, because a history of selection has favoured individuals that adopted them.

Enthusiasts of the meme idea frequently take recourse in analogy with the virus. Perhaps because of the need to demonstrate that culture is a genuine evolutionary process in its own right and cannot be reduced to a mere product of biological evolution, so far meme enthusiasts have concentrated almost exclusively on the characteristics that make memes infectious. However, the success of a virus depends not only on its *infectiousness* but also on the *susceptibility* of its hosts and on whether the *social environment* promotes contact between hosts. The same three factors may also determine the success of memes (Laland and Odling-Smee, 2000). Were memeticists to accept that evolved genetic predispositions may influence meme adoption, leaving human beings particularly susceptible to acquiring memes that increase their reproductive success, they would converge on the ideological position of advocates of gene–culture coevolution. We will return to this debate in the two remaining chapters. However, one point is worth stressing here. To acknowledge that genetic predispositions may have some role to play in cultural change is not tantamount to the suggestion that transmitted culture is irrelevant, or that genes completely determine which memes are adopted. While cultural processes require explanation in their own right, they are not entirely divorced from biology (Cavalli-Sforza and Feldman, 1981; Boyd and Richerson, 1985).

Could there be a science of memetics?

If there is one criticism of memetics that has genuine merit, it is that, while memes offer an interesting alternative panorama, memetics is not yet a science. Memetics is a social club in which Dawkins and Dennett fans put on their meme's-eye view goggles and entertain each other with fanciful evolutionary stories. But could that change? Is a science of memetics possible? Dennett poses the question:

> Philosophers, some will say, may appreciate the (apparent) insight to be found in a striking new perspective, but if you

can't turn it into actual science, with testable hypotheses, reliable formalizations, and quantifiable results, what good is it, really? (1995, p. 353)

Dennett's answer is surprisingly cautious:

The prospects for elaborating a rigorous science of memetics are doubtful, but the concept provides a valuable perspective from which to investigate the complex relationship between cultural and genetic heritage. (1995, p. 369)

Dennett's reservations stem from the apparent difficulties of doing empirical work on meme evolution. It is not clear whether we will ever be able to decipher the informational content of networks of neurons, so that, at least for the foreseeable future, we can't read the memes in a brain in the same way we can read the genes on a chromosome (but see Aunger, 2002). Moreover, for Dennett, memes mutate so frequently and meme phenotypes change so quickly that memes simply can't be tracked and the prospects for 'cranking out a science that charts ... [meme] descent are slim' (Dennett, 1995, p. 356). However, as Aunger (2000a) points out, all of Dennett's reservations are untested empirical assertions that may not be true. To be fair to memeticists, it is a new discipline and it does lay claim to a handful of empirical studies (Hull, 2000). Moreover, recent years have seen the emergence of theoretical methods for testing hypotheses about memes, which show considerable promise (Pocklington and Best, 1997; Kendal and Laland, 2000; Bull et al., 2001).

Nonetheless, ultimately memetics will stand or fall on whether it generates empirical research (Aunger, 2000a; b). Meme advocates must accept that, unless they devise a rigorous methodology for *doing* memetics, methods that instigate a valid research programme involving testing as

234 SENSE AND NONSENSE

well as generating hypotheses, then memetics will never be a science.

Harsh critics even dismiss the whole enterprise of memetics as fundamentally tautological (Wilson, 1999). If enthusiasts do little more than conjure up stories about how only the fittest memes survive, and then simply assume that memes are fit because they've survived, this criticism is entirely justified. Were the same charge to be levelled at population geneticists concerning biological evolution they could counter that natural selection has been clearly demonstrated in countless natural populations by utilizing a variety of experimental methods (Endler, 1986a). If memeticists are correct in their claim that natural selection operates on cultural variation then they should be able to detect that selection using parallel methods. If anything, applying such tools to testing natural selection among memes should be easier than it is for genes, because of the greater rates of change that are typically found. What are these methods?

In his book *Natural Selection in the Wild* (1986a), evolutionary biologist John Endler describes ten experimental methods successfully used to detect natural selection. Equivalent procedures for memetics can be devised in all cases. Here we present simplified versions of just five of these methods translated into the language of memetics, but we recommend that potential practitioners return to Endler's original text for further details, qualifications, and caveats. Methods for detecting the natural selection of cultural variation include the following.

1. *Correlations between selected and selecting factors.* If natural selection occurs, then geographical variation in the selective factor will give rise to parallel variation in traits. In the case of memes, the selective environments are human

minds and prior knowledge, including other memes. Thus this method boils down to making *a priori* predictions concerning correlations between memes and then testing whether such correlations are found. It is important that the 'selected' memes spread independently of the 'selecting' memes, and not simultaneously.

2. *Cultural character displacement.* Where two closely related conceptual lineages compete, their characteristics should diverge, reducing competition (character displacement). Researchers could identify two homologous conceptual systems (e.g. protestantism versus catholicism), and test whether they are more different in regions where they come into contact than where they do not.

3. *Convergent conceptual evolution.* Researchers could make comparisons between unrelated conceptual systems that thrive in similar selective environments, predicting that the same or similar memeplexes will evolve. For example, is there convergence in technological innovation? Are there examples of the bow and arrow, the arch, or penicillin where different populations confronted with the same problem independently arrive at the same solution? Perhaps independent information sources, such as historical records, can be used to verify these patterns.

4. *Perturbations of natural selection.* Meme frequency distributions are unlikely to be at equilibrium immediately after a natural or artificial change in their prevalence. If, after the pulse, meme frequencies change more than expected through chance, and in a consistent direction, this is evidence for cultural evolution. For example, researchers could investigate whether and how food sales recover from a food hygiene scare, tourism rejuvenates after a foot-and-mouth outbreak, or luxury goods come back into fashion following a war or recession.

5. *Optimality models*. Researchers could predict equilibrium distributions of memes on the basis of known properties of meme phenotypes using optimization models. The model would need to identify a priori the most successful strategy for passing on memes given relevant constraints, which could then be tested in the real world. This would involve identifying who is a likely host for the meme, how they are best contacted, what are the most effective methods for persuasion, etc. Such an approach could be applied to data from politics or advertising.

Strictly, the above methods will not distinguish natural selection among memes from other forms of cultural evolution (see Boyd and Richerson, 1985; also, see Chapter 7), or from the acquisition of cultural traits through interaction with artefacts. However, any young science has to start at the beginning. As Aunger says:

> what we really want to know is whether selection is
> directional rather than neutral, and to identify the selective
> agent. The answers to these kinds of questions can get us a
> long way toward an understanding of the evolution of the
> system under study and may be possible for a future
> memetics. (2000b, p. 14)

There are other contributions that an empirical approach to memetics could make. Hull (2000) suggests that memeticists carry out experiments that improve our understanding of the mechanisms involved in memetic transmission. Postulating memes for this and that risks a circulation of reasoning, if memes are not only 'what we think' but also the *explanation* for what we think. There is a real need to investigate the developmental processes that lead to individuals acquiring the beliefs that they hold. This would involve not only experimental studies of the psychological processes that underpin social learning, but also studies of the processes of social interaction that facilitate or impede meme transmission. Other experiments, perhaps using *transmission chains*,[2] could explore to what extent there are isolatable cultural units and, if there are, to what extent they are characterized by longevity, fecundity, and fidelity. For instance, if celibacy has evolved for the sake of the memes, researchers should be able to test whether the parishioners of catholic priests have acquired more of the denomination's message than those of protestant priests. Researchers could investigate experimentally to what extent socially learned information is transmitted or reconstructed, and whether reconstruction requires the relevant genotype. Moreover, virtually all such experiments could be carried out using humans or other animals.

> Memeticists cannot begin to understand what the science of memetics is until they generate some general beliefs about conceptual change and try to test them. These tests are likely to look fairly paltry, but in the early stages of a science, attempts at testing always look paltry. (Hull, 2000, p. 4)

Conclusions

In our judgement, memes do have value as a tool for thought that challenges us to consider the possibility that some human characteristics may exist or become prevalent purely by virtue of their ease of propagation. Potentially, memes provide a pragmatic means of quantifying and

---

[2] Transmission chains investigate the stability of socially transmitted information by examining whether it can spread from one individual to the next, or one group to the next, in a chain (see Laland, 1999 for a more extensive description of the methods).

operationalizing cultural phenomena and describing cultural change, in the process lending them more readily to scientific enquiry. Whether memetics provides the comprehensive theory of human nature that some of its most passionate devotees claim is perhaps more contentious.

Michael Turelli, one of the world's leading evolutionary biologists, used to quip 'who needs babies when you can have reprints!' While scientists derive considerable satisfaction when professional colleagues understand and accept their theories, few would trade in the more conventional form of reproduction for a conceptual legacy. Yet, if Dawkins is correct, when human beings die there are two things that are left behind: genes and memes. Our genes may be immortal but the collection of genes in each of us disperses. However,

> if you contribute to the world's culture, if you have a good
> idea, compose a tune, invent a sparking plug, write a poem,
> it may live on, intact, long after your genes have dissolved
> into the common pool ... The meme-complexes of
> Socrates, Leonardo, Copernicus and Marconi are still going
> strong. (Dawkins, 1976, p. 214)

Will the 'meme' meme long outlive the dissolution of Dawkins' genes? The answer to this question will largely depend on whether memetics can become a progressive research programme (Aunger, 2000a). One means by which it might become more than a flash-in-the-pan is if a serious science of population memetics could be derived, which develops formal theoretical models that predict the patterns and rates of cultural evolution in the same way that population geneticists predict the characteristics of biological evolution. A second factor that may well determine its fate is whether meme enthusiasts will take seriously the view, advocated by virtually all of the other evolutionary approaches described in this book, that human beings have brains that have evolved to favour some ideas, beliefs, and knowledge over others. In the next chapter we will see that something very close to a theoretical foundation for memetics already exists, an approach that allows both biological and cultural evolution to be explored simultaneously. That approach is called gene–culture coevolution.

CHAPTER 7

Gene–culture coevolution

For most human behavioural ecologists and evolutionary psychologists, the cardinal aspects of human nature reflect the imperatives of genes and the environments in which they are expressed. In stark contrast, memeticists believe that cultural processes provide a more powerful explanation for the interesting aspects of our conduct, a belief that they share with most social scientists. While the parties wrangle over the relative significance of genes and culture, in truth virtually everyone accepts that both are important. But is it possible to do better than this? Could we specify *how* genes and culture interact, together with what role other factors in the environment play? If genes and culture both evolve, what is to stop them adapting to each other or modifying the other's selective environment? Could genes and memes even be wrestling with each other for control of their own destinies?

Stone tools appear in the archaeological record approximately two and a half million years ago. The significance of this observation is not simply that *Homo habilis* and later hominid species had the guile to manufacture a lithic technology, but also that these skills were transmitted from one generation to the next. These simple artefacts thus represent the earliest evidence for culture. In fact, comparative evidence for social learning in a variety of vertebrate species suggests that cultural transmission appeared long before the advent of our genus. However, social learning in other animals is rarely stable enough to support traditions in which significant amounts of information accumulate from one generation to the next. For at least two million years, our ancestors have reliably inherited two kinds of information, one encoded by genes, the other by culture. How does this dual inheritance affect the evolutionary process?

There is only one evolutionary approach to the study of human behaviour that takes up the challenge of understanding genetic and cultural evolution simultaneously by focusing directly on their interaction. This is the third of the three principal evolutionary approaches that emerged in the aftermath of the human sociobiology debate (Smith, 2000). It goes by the names of 'gene–culture coevolutionary theory' or 'dual-inheritance theory', the first term having been coined by Stanford geneticists Marc Feldman and Luca Cavalli-Sforza, and the second by UCLA and UC Davis anthropologists Robert Boyd and Peter Richerson. Although some observers have characterized these labels as representing different conceptions of the relationship between genes, development, and culture (e.g. Flinn, 1997), most researchers within the field see no difference in perspective and regard the terms as synonyms. Research within this tradition is also sometimes described as 'cultural evolution' or 'cultural selection', but we avoid these terms as they apply more to analyses of a single (cultural) inheritance process, rather than two (genetic and cultural) interacting forms of inheritance. We will use the term 'gene–culture coevolution' from now on to describe this approach.

Gene–culture coevolution is like a hybrid cross between memetics and evolutionary psychology, with a little mathematical rigour thrown into the pot. Like memeticists, gene–culture coevolution enthusiasts treat culture as an evolving pool of ideas, beliefs, values, and knowledge that is learned and socially transmitted between individuals. Like evolutionary psychologists, these researchers believe that the cultural knowledge an individual adopts may sometimes, although certainly not always, depend on his or her genetic constitution. For instance, if Fred lacks the genes for alcohol tolerance he is unlikely to develop a taste for single-malt whisky. Moreover, selection acting on the genetic system is commonly generated or modified by the spread of cultural information. Below we describe how a history of dairy farming created the selection pressures that favoured the genes underlying the capacity of adult humans to consume dairy products. For gene–culture coevolutionary theorists, the 'leash' that ties culture to genes tugs both ways. The advent of culture was a precipitating evolutionary milestone, generating selection that favoured a reorganization of the human brain that left it specialized to acquire, store, and utilize cultural information. It was culture, loosely guided by genes, that allowed humans the adaptive flexibility to colonize the world.

The quantitative study of gene–culture coevolution began in 1976, when two of the world's leading geneticists, Marcus Feldman and Luca Cavalli-Sforza, published the first simple dynamic models with both genetic and cultural inheritance. The innovative aspect of their work was that, in addition to modelling the differential transmission of genes from one generation to the next, they incorporated cultural information into the analysis, allowing the evolution of the two systems to be mutually dependent. However, one curious feature of the history of gene–culture coevolution is that both archetypal sociobiologists and some of their most severe critics almost simultaneously recognized the importance of gene–culture interactions, with each starting to develop methods to address the problem. By the late 1970s, Charles Lumsden and Edward Wilson were self-explicitly engaged in a race with Cavalli-Sforza and Feldman to produce the first book on this topic (Segerstråle, 1986). While Lumsden and Wilson's (1981) *Genes, Mind and Culture* was published first, it was to receive an unfriendly reception. In contrast, Cavalli-Sforza and Feldman's (1981) more cautious tome *Cultural Transmission and Evolution* was much better received.

Lumsden and Wilson described gene–culture coevolution as:

> a complicated, fascinating interaction in which culture is generated and shaped by biological imperatives while biological traits are simultaneously altered by genetic evolution in response to cultural innovation. (1981, p. 1)

For mathematical convenience, and like many subsequent workers, they assumed that culture could be learned as discrete packages, or 'culturgens': units of culture synonymous with Dawkins's memes. However, for Lumsden and Wilson, an individual's choice of culturgen was affected by a combination of 'genetically determined epigenetic rules' and social learning. The models that they produced allowed Lumsden and Wilson to predict how the cultural information and the genes underlying the epigenetic rules would change over time and across different cultures. This led to a number of conclusions: for instance, they found that it is extremely likely that genetic biases would evolve that affect what cultural information is adopted, that weak genetic biases can be amplified by conformity of behaviour and have a major influence on the characteristics of populations, and that culture can both slow down and speed up the rate of genetic change.

Perhaps unsurprisingly, Lumsden and Wilson's book received severe criticism, with many readers suspicious that the models' assumptions stacked the deck in favour of the genetic control of culture (Maynard-Smith and Warren, 1982; Lewontin, 1983b; Boyd and Richerson, 1983; Kitcher, 1985). However, published as it was at the height of the sociobiology debate, Lumsden and Wilson's mathematical treatise was never likely to be judged completely objectively and as a consequence assessment of it was heavily influenced by the views of a small number of hostile reviewers (see Chapter 3).

In contrast, the work of Feldman and Cavalli-Sforza was to have a more lasting influence. Along with a substantial number of co-workers, Feldman and Cavalli-Sforza gradually built up an impressive body of mathematical theory exploring the processes of cultural change and interaction between genes and culture. They frequently took advantage of the parallels between the spread of a gene and the diffusion of a cultural innovation to borrow or adapt established models from population genetics. These researchers largely disavowed Lumsden and Wilson's work, and sometimes challenged their findings. Together, Feldman and Cavalli-Sforza laid the theoretical foundations for an entire new field.

Fuelled by the ongoing sociobiology debate, other mathematically minded researchers joined the fray, most notably anthropologists Robert Boyd and Peter Richerson, whose *Culture and the Evolutionary Process* (1985) introduced a variety of novel theoretical methods and stimulating ideas. Boyd's flair for mathematical modelling combined with Richerson's encyclopaedic knowledge were to prove an irresistible combination, and their book, for which they were awarded the prestigious Staley prize from the School of American Research, won them many plaudits. Gradually a consensus as to the most appropriate methods for tackling gene–culture interactions began to emerge, which today forms the basis of modern gene–culture coevolutionary theory.

The highly technical and explicitly mathematical nature of modern gene–culture coevolution is one of several features that distinguishes this perspective from the alternatives. A second is its explicitly non-adaptationist stance, by which we mean the incorporation into the analyses of a variety of genetic and cultural processes in addition to the natural selection of genes. Gene–culture coevolution exhibits a concern for non-adaptive and even maladaptive outcomes of the evolutionary process. This stance continues both to surprise and confuse outside observers more used to characterizing all these evolutionary approaches as 'sociobiology'. However, the rigorous theoretical approach has not led to many experiments or other forms of empirical work and this school remains the prerogative of a comparatively small band of workers whose research frequently exhibits a rather esoteric flavour.

Nonetheless, the emerging body of theory has developed in a variety of ways. One class of models investigates the inheritance of behavioural and personality traits (Cavalli-Sforza and Feldman, 1973; Otto et al., 1995). Other models explore the adaptive advantages of learning and culture (Rogers, 1988; Boyd and Richerson, 1985; Feldman et al., 1996). More recently, these methods have been applied to address specific cases in which there is an interaction between cultural knowledge and genetic variation that influences its prevalence. These include the evolution of language (Aoki and Feldman, 1987; 1989) and handedness

(Laland *et al.*, 1995b), an analysis of changes in the genetic sex ratio in the face of sex-biased parental investment (Kumm *et al.*, 1994), the spread of agriculture (Aoki *et al.*, 1996), the coevolution of hereditary deafness and sign language (Aoki and Feldman, 1991), the emergence of incest taboos (Aoki and Feldman, 1997), and an exploration of how cultural niche construction affected human evolution (Laland *et al.*, 2001).

The problems addressed by gene–culture coevolution are of fundamental interest to the biological and social sciences. Do our genes restrict and delineate the nature of our culture? What processes underlie human cooperation and conflict? How did culture evolve, and how has it affected evolution in our lineage? Perhaps the simultaneous focus on biological and cultural processes will give gene–culture coevolution an advantage in the quest to understand human behaviour. This chapter provides a guided tour to gene–culture coevolutionary models, describing them in simple, non-mathematical terms, explaining the aims and assumptions of the modellers, and critically analysing their methods and conclusions.

Key concepts

In this section, we address the importance of cultural inheritance in gene–culture coevolutionary theory. We then describe some of the processes by which particular cultural phenomena may change in frequency and the routes that the transmission of information might take. Finally, we use a non-technical example to illustrate how gene–culture models work, and discuss the insights and limitations that can come from building models.

Cultural inheritance

For most social scientists 'culture' is a given. The notion
that much of the variation in the behaviour of humans is
brought about by their being exposed to divergent cultures
is so widespread and intuitive that it appears beyond dispute. Culture is regarded as a cohesive set of mental representations, a collection of ideas, beliefs, and values that are
transmitted among individuals and acquired through social
learning.

In contrast, most sociobiologists and evolutionary psychologists are united by the belief that the transmitted
elements of culture exert either a comparatively trivial
influence on human behaviour, or that whatever influence
they have is so strictly circumscribed by genes that there is
no need to take account of the dynamic properties of culture. For human behavioural ecologists, culture is viewed as
a flexible system that produces the most adaptive outcome
in a given environment and that can be altered over a relatively short period of time in response to environmental
change. Others, such as many behaviour geneticists, treat
'culture' as the dross that is left over when the 'more important' genetic influences on behaviour have been isolated.
'Culture' is usually lumped together with individual learning and other environmental effects on behaviour into a
ragbag labelled 'nurture', to be contrasted with genetic
sources of variation.

For proponents of gene–culture coevolution, many of
these other biological perspectives are misguided. Too
much culture changes too quickly to be feasibly explained
by genes, while the fact that different behavioural traditions can be found in similar environments would
appear to render environmental explanations of behaviour
impotent a lot of the time (Boyd and Richerson, 1985). To

give an example, Guglielmino *et al.* (1995) carried out an analysis of variation in cultural traits among 277 contemporary African societies and found that most traits examined correlated with cultural history rather than with ecology. Such findings suggest that most human behavioural traits are maintained in populations as distinct cultural traditions rather than being evoked by the natural environment. Genes and environment undoubtedly account for some variation in human behaviour but the socially transmitted component of culture is hard to ignore.

Our capacity for culture is a unique adaptation. It allows us humans to learn about our world rapidly and efficiently. Human beings don't have to scour their environment for sources of food and water, devise their own means of communication, or reinvent technological advances from first principles. Our capacity to acquire valuable skills and information from more knowledgeable others, such as parents, teachers, or friends, as well as indirectly via artefacts such as books and computers, furnishes us with a short cut to adaptive (and sometimes maladaptive) behaviour. Advocates of gene–culture coevolution share with memeticists and the vast majority of social scientists the view that what makes culture different from other aspects of the environment is the knowledge passed between individuals. Culture is transmitted and inherited in an endless chain, frequently adapted and modified to produce cumulative evolutionary change. This infectious, information-based property of transmission is what allows culture to change rapidly, to propagate a novel behaviour through a population, to modify the selection pressures acting on genes, and to exert such a powerful influence on our behavioural development.

Gene–culture enthusiasts point to countless studies that have found that the attitudes of parents and offspring are rather similar, and maintain that the most obvious explanation for this is that children learn social attitudes in the family. For instance, a study of Stanford University students revealed that the religious and political attitudes were strongly consistent between parents and offspring (Cavalli-Sforza et al., 1982). The same has been reported to apply in non-industrial societies. Among Aka pygmies, an African group of hunter–gatherers, there was evidence for parent to child transmission of many customs (Hewlett and Cavalli-Sforza, 1986), while among horticulturists in the Democratic Republic of Congo the young acquire knowledge about foods primarily from their parents (Aunger, 2000c). Such correlations do not prove cultural transmission to be prevalent, but the weight of evidence supports the notion of a transmitted culture (see Boyd and Richerson, 1985, for a more extensive collation of evidence for cultural transmission).

Types of cultural selection

In the previous chapter, we saw how meme enthusiasts suggest that culture can evolve in its own right, but they rarely specify in any detail exactly how memes change in frequency. By contrast, researchers in the gene–culture coevolution tradition have described a number of processes that underpin cultural change. To distinguish cultural from biological evolution, Cavalli-Sforza and Feldman (1981) define *cultural selection* as a process by which particular socially learned beliefs, or pieces of knowledge, increase or decrease in frequency due to being adopted by other individuals at different rates. Meanwhile, *natural selection* can also change the frequency of a cultural preference, through the differential survival of individuals expressing different types of preference. For instance, in developed countries, fertility control (e.g. via contraception) is at a clear disadvantage in terms of natural selection as users typically have fewer offspring, but has spread by virtue of its advantage in cultural selection since fertility control is a popular choice. Working with biological and cultural processes simultaneously could help us to understand how nonadaptive cultural traditions might evolve (Cavalli-Sforza and Feldman, 1981). When it has sufficiently high cultural fitness, cultural information could increase in frequency despite decreasing genetic fitness.

The cultural traditions of a population may change over time if individuals alter the cultural information that they receive before passing it on. Boyd and Richerson (1985) discuss *guided variation*, which refers to a process by which individuals acquire from others information about a behaviour, and then modify the behaviour on the basis of their personal experience. Here cultural variation is guided by individual experience, which may allow behavioural traditions to evolve gradually towards the optimal behaviour for that environment, as human behavioural ecologists envisage.

Given a choice been two alternative behaviour patterns, individuals may be more likely to adopt one variant than another (Cavalli-Sforza and Feldman, 1981). Boyd and Richerson (1985) refer to this as *biased cultural transmission*. Various types of bias may exist. In *direct bias*, individuals choose which of two or more alternative behaviour patterns to adopt. A direct bias might result from a genetic predisposition to favour certain types of information, similar to Lumsden and Wilson's epigenetic rules or to evolutionary psychologists' conception of learning. Stanford University anthropologist Bill Durham (1991) has argued that the individual choices that underpin these cultural processes are guided, but not determined, by predispositions and prior knowledge.[1] As sociobiologists and evolutionary psychologists envisage, genetically biased transmission is likely to generate adaptive behaviour much of the time (Boyd and Richerson, 1985).

The cultural traditions that an individual picks up will often depend upon who else in the population has adopted that tradition. In the case of *frequency-dependent bias*, the commonness or rarity of a behaviour affects the probability of information transmission. When, as often seems to be the case, individuals are predisposed to adopt the behaviour of the majority, this frequency-dependent bias generates *conformity*. We shall see that conformist transmission has some interesting consequences; for instance, it could result in a viable form of group selection and it can lead to maladaptive outcomes. People may also use cues about one trait, for example wealth, to choose which individuals to observe in order to acquire information about another trait, such as clothes' fashions. This form of learning is called *indirect bias* by Boyd and Richerson.

Researchers in gene–culture coevolution are also interested in how information spreads within populations. The

---

[1] Durham (1991) suggested that our decisions are guided by *primary values*, which represent the valuative feedback produced by our nervous system and informed by our genes, and *secondary values*, which represent our previously acquired socially transmitted knowledge.

mode of transmission describes the route by which cultural knowledge passes among individuals (Cavalli-Sforza and Feldman, 1981), and different models are required for alternative modes of information transmission. Social transmission can occur *vertically* (that is, from parents to offspring), *obliquely* (from the parental to the offspring generation; for instance, learning from teachers or religious elders) or *horizontally* (that is within-generation transmission, such as learning from friends or siblings). Of course, genetic inheritance is exclusively vertical and hence, as social transmission frequently occurs through some combination of these modes of information transmission, cultural evolution and gene–culture coevolution may commonly exhibit quite different properties from biological evolution.

Constructing gene–culture models

The construction of gene–culture models is a complicated procedure, and it would be beyond the scope of this book to give a detailed account of this process. Instead we will illustrate the logic behind the gene–culture coevolutionary method with a deliberately simplified example that illustrates how researchers describe changes in the frequency of a meme in a population that are brought about by cultural and natural selection.[2]

Imagine ten friends, six male (Bob, Jim, Harry, Bert, Ted, and Hank) and four female (Jenny, Jean, Sally, and Sue).

---

[2] For a more technical introduction see Feldman and Cavalli-Sforza (1976) or Laland *et al.* (1995a).

Several of these friends share a hobby of racing cars. We will assume for simplicity that individuals can be categorized as *racers*, who love the adrenaline rush of high-speed driving but who are sometimes reckless and will take chances to win a race, and *non-racers*, who regard fast driving as frightening and prefer not to take part. Five of the friends are racers (Bob, Jim, Bert, Hank, and Sally), while the others are not. Now what if Harry, under severe pressure from his male friends and goaded for his wimpiness by girlfriend Sally, takes up racing? How could we track this change in racing behaviour within the population? For the moment, ignore any possible influence of an individual's sex on the propensity to be a racer. We know that prior to Harry's switch the proportion of males among the friends was six out of ten (that is, 0.6) and the proportion of racers was five out of ten (that is, 0.5). From this, we would expect a proportion $0.6 \times 0.5 = 0.3$ of the population, or three of the ten individuals, to be racing males. However, this was not the case, as four of the males were racers. By similar reasoning, after Harry has taken up racing we would expect $0.6 \times 0.6 = 0.36$, that is, between three and four racing males; however, there are in fact now five racing males. There seems to be a discrepancy between the expected and observed numbers of each sex that are racers or non-racers. What is wrong here?

The errors occur because of a non-random association between genes (that is, sex) and memes (racing or not). There is a departure from what we would expect by chance because a disproportionate number of males are racers. To describe the pattern among these friends accurately, we can't just generalize from the overall proportions of each sex and each meme in the group. Instead, we would have to track separately the gene–meme combinations, or the proportions of racing males, racing females, non-racing males, and non-racing females.[3]

The frequencies of each gene–meme combinations could be affected by two types of process. There are cultural selection processes, such as Harry's conversion, an example of horizontal cultural transmission that changed the proportion of racing males from 0.4 to 0.5. There are also natural selection processes in operation. For instance, imagine that, while racing, Hank tragically dies in a terrible accident, which drops the proportion of racing males back down to four out of nine, or 0.44. However, Ted and Jean become a couple and give birth to a baby girl who, initially at least, is a non-racer, which increases the proportion of non-racing females in the group from 0.33 to 0.4. It would be relatively easy to construct mathematical expressions that describe the proportion of each gene–meme combination in the group each year, as a function of what they were in the previous year, and how the frequencies have been changed by conversions to or from racing as well as birth and death processes.

The above example illustrates the logic of a gene–culture coevolutionary analysis. While the change in frequency of genes and cultural knowledge can sometimes be modelled separately, in other cases, because of interactions, we need to keep track of the change in frequency of gene–meme

---

[3] The rather cumbersome technical name given to a particular combination of genotype and meme is the *phenogenotype* (Feldman and Cavalli-Sforza, 1976). It is the smallest non-divisible unit of gene–culture coevolution that can be counted. Non-random association between genes and memes is known as *phenogenotype disequilibrium*. An extensive analysis of such interactions can be found in Feldman and Zhivotovsky (1992).

combinations. In either case, in addition to the rules of Mendelian inheritance,[4] transmission rules for cultural information must be described. Ways to formalize such rules have been developed by Cavalli-Sforza and Feldman (1981) and others. The most common assumption is that the probability of an individual adopting a belief or preference depends on whether his or her parents have that belief, but equivalent models have been developed in which learning is either from unrelated individuals, from key individuals in the social group or from the majority in the group.

A set of such transmission rules is depicted in Table 7.1. Remaining with our racing example, we assume that over the years the friends pair up and have children. What are the probabilities that they will grow up to be racers or non-racers? If children are influenced by their parental teachings then whether or not a particular child becomes a racer will depend, in part, on whether its parents were racers. We can set parameters that represent these probabilities of vertical cultural transmission, which are the b_i terms in Table 7.1.

Table 7.1 Transmission rules for the probability of children becoming racers or non-racers depending upon the racing behaviour of the parents

Mating type		Probability of	
Mother	Father	Racing child	Non-racing child
Racer	Racer	b_3	$1-b_3$
Racer	Non-racer	b_2	$1-b_2$
Non-racer	Racer	b_1	$1-b_1$
Non-racer	Non-racer	b_0	$1-b_0$

---

[4] Strictly, gene–culture coevolutionary analyses are further complicated by the fact that, unlike for sex, the genetic component is heritable.

Let us assume that children with two racing parents are more likely to develop an enthusiasm for racing than children with just one racing parent, who in turn are more likely to become racers than children with two non-racing parents. We can represent this in the analysis by simply setting the b_i parameters such that b_3 is greater than b_2 and b_1, both of which are greater than b_0. To give a specific example, this would occur if the children of two racing parents always become racers (so that $b_3 = 1$), children of two non-racing parents never become racers ($b_0 = 0$), and children with one racing parent become racers half of the time ($b_1 = b_2 = 0.5$). This case represents unbiased vertical cultural transmission and, if such rules apply, there would be no overall change in the frequency of racing behaviour as a consequence of cultural processes, although the frequency of racing might decrease as a consequence of natural selection when the odd driving accident occurs.

However, what if driving fast is so thrilling that racing parents can't stop talking about it? This might create a transmission bias, making it slightly more likely that children with one racing parent would become racers than non-racers (b_1, $b_2 > 0.5$). Now there are two conflicting processes that act on the frequency of racing: cultural selection favours racing and acts to increase its frequency, while natural selection favours individuals that don't race. Depending on the relative strengths of these two processes, racing may or may not increase in frequency.

Lastly, consider Daly and Wilson's (1983) hypothesis that the higher level of road accidents among males is a manifestation of a history of sexual selection in which human males were selected for risk-taking strategies, while females were selected to be more risk-averse. If that is the case, then the probability that a child becomes a racer depends not only on its parents' memes, but also on its own genes (that is, on its sex). Now the b_i parameters will take on different values depending on whether the child is male or female, and the frequency of racing will differ among the sexes, being higher in males. Even if a genetic predisposition toward racing is found among sons (b_{1m}, $b_{2m} > 0.5$) but not daughters ($b_{1f} = b_{2f} = 0.5$), the frequency of racing will reach higher than chance levels in females as well as in males. This is because there will be an increasing number of families in which at least one parent will be a racer, which will influence the chances of daughters as well as sons becoming racers.[5]

Cultural transmission procedures such as those in Table 7.1, combined with rules for mating and genetic inheritance, allow gene–culture coevolutionary researchers to derive a system of equations that describes how the relevant gene and meme frequencies change over time in the face of cultural selection, natural selection, and various kinds of interactions and biases. In our example, the individuals in the group either did or did not exhibit the behaviour pattern (they were either racers or non-racers). However, equivalent sets of equations can also be produced where the behaviour of an individual can be placed somewhere along a scale; for example, where individuals are described according to the average speed at which they drive. The value of constructing gene–culture coevolutionary models

---

[5] Note that while several gene–culture models have incorporated the assumption that an individual's genotype influences the probability that particular cultural information will be adopted, practitioners are also free to assume that information may be adopted independent of their fitness consequences. Researchers can set up a model based on reasonable assumptions and see if the behaviour that results is adaptive or not.

is that it allows researchers to ask questions such as, 'Could a predisposition for risk-prone behaviour be favoured in males?', 'Under what circumstances can memes (such as driving fast) spread even if they reduce Darwinian fitness?', 'What will be the final frequency of the (racing) behaviour in the population, when it reaches equilibrium?', and 'How much of the variability in peoples' (driving) behaviour can be attributed to differences in genes, differences in parental behaviour (vertical cultural transmission), alternative social influences (horizontal and oblique transmission) or other factors?' The benefits of such modelling are twofold: first, mathematical models provide researchers with an understanding of processes that cannot be studied in other ways. For instance, comparatively little is known about human evolution, and researchers can't carry out selection or breeding experiments using humans to test hypotheses about our evolutionary past. They can, however, develop mathematical models of such processes, analyse them, and use the results to test the feasibility of their hypotheses. Second, as we saw in the human behavioural ecology chapter, models can be a useful guide to empirical research. For instance, models frequently generate testable predictions, and highlight the key factors that researchers need to measure. In this manner, mathematical analyses can be evaluated with empirical data.

Case studies

In this section, we present three examples of research carried out using the methods of gene–culture coevolution. These include an investigation of the coevolution of the cultural practise of dairy farming and the genes that allow adult humans to digest milk, a model of the group selection of cultural variation, and an analysis of the factors that explain why people have different levels of intelligence.

Coevolution of dairy farming and genes for processing milk

The evolution of the ability of adult humans to consume dairy products represents a good example of gene–culture coevolution. Unlike human infants, virtually all of whom can drink cows' milk without problems, adult humans vary considerably in their ability to digest milk as a result of differences in their physiology. In fact, if the entire world's population is considered, consuming dairy products actually makes the majority of adult humans ill. This is because the activity level of the enzyme lactase in their bodies is insufficient to break down the energy-rich sugar lactose in dairy products, and milk consumption typically leads to sickness and diarrhoea. Whether or not adult humans can digest lactose is largely down to whether they possess the appropriate copy (or allele) of a single gene. It turns out that a strong correlation exists between the incidence of the genes for lactose absorption and a history of dairy farming in populations, with absorbers reaching frequencies of over 90% in dairy farming populations but typically less than 20% in populations without dairy traditions (Simoons, 1969; Durham, 1991). Milk and milk products have been a component of the diets of some human populations for over 6,000 years, roughly 300 generations. Is it conceivable that dairy farming created the selection pressures that led to the allele for absorption becoming common in pastoralist communities? Gene–culture coevolutionary theory is precisely the kind of analysis that can answer this question.

Following work by Aoki (1986), Feldman and Cavalli-Sforza (1989) used gene–culture coevolutionary models to investigate the evolution of lactose absorption. They constructed a model in which the capacity to absorb lactose was affected by alleles of a single gene, with one particular allele allowing adults to digest milk without getting sick, and in which milk usage was a tradition learned from other members of the population. Their model showed that whether or not the allele allowing adult milk digestion achieved a high frequency depended critically on the probability that the children of dairy product users themselves became milk consumers (equivalent to the b_3 parameter in Table 7.1 if dairy product users are substituted for racers). If this probability was very high then a significant fitness advantage to the genetic capacity for lactose absorption resulted in the selection of the absorption allele to high frequency within 300 generations. However, if a significant proportion of the offspring of milk users did not exploit dairy products then unrealistically strong selection favouring absorbers was required for the gene for absorption to spread. In other words, differences in the strength of cultural transmission between cultures may account for genetic variability in lactose absorption. Thus the analysis is able to account for both the spread of lactose absorption and the culturally related variability in its incidence. Moreover, there were a broad range of conditions under which the absorption allele did not spread despite a significant fitness advantage, indicating that traditional genetic models would frequently get the wrong answer. Cultural processes complicate the selection process to the extent that the outcome may differ from that expected under purely genetic transmission.

Over recent years the dominant view among the scientific community has been that adult lactose tolerance in humans is an adaptation to reduced exposure to the sun: both the sun and the enzyme lactase promote calcium absorption so, in populations living at high latitudes, lactose production may have been selected as an alternative method of absorbing calcium (Durham, 1991). However, Holden and Mace (1997) applied recently developed statistical methods (Pagel, 1994) to a phylogeny of human cultural groups and found no evidence for the latitudinal theory but strong support for the dairy farming hypothesis. In addition, their analysis suggested that dairy farming evolved first, which then favoured tolerance to lactose, and not the other way around. Holden and Mace's analysis provided compelling confirmation of the findings of the gene–culture co-evolutionary model.

Cultural group selection

Over the years, one of the most hotly debated topics within evolutionary biology has been whether natural selection can operate on groups of individuals. If group selection occurs, it could result in characteristics that evolve for the good of the population; for instance, behaviour and institutions that promote altruism and cooperation. Most evolutionary biologists accept the findings of theoretical models suggesting that group selection is plausible but only under restricted conditions (Price, 1970; Uyenoyama and Feldman, 1980) and many question how frequently such conditions arise naturally. For example, one of the requirements for selection at the level of the group is that genetic differences between groups are maintained. However, the processes that uphold group differences and select between groups are typically weak compared with the processes that break down group differences and select within groups (Williams, 1966; Dawkins, 1976). For instance, genetic differences typically arise through 'genetic drift' (that is, random changes in the genetic composition of the group), but movement of individuals between groups will quickly erode these differences. Another problem is that any group that exhibits cooperation between individuals will be susceptible to individuals who cheat and gain the benefit without paying the costs, and these cheats are expected to thrive. For group selection to be operational, groups of altruists would have to give rise to new altruistic groups significantly more frequently, or go extinct less frequently, than groups without altruists while somehow counteracting the influence of selfish interest. As we saw in Chapter 3, the sociobiological revolution was built upon a rejection of group selection, and many biologists regard the group selection of genetic variation to be insufficiently strong to counter the eroding action of natural selection within groups (although see Sober and Wilson, 1998, for a counter position).

Boyd and Richerson (1982; 1985) propose an alternative form of group selection that might just work. Their hypothesis stresses the group selection of cultural rather than genetic variation, a process that surmounts many of these problems.[6] Many social scientists believe that people conform to the social norms of their society without much thought. Rather than working out how to behave from scratch, a lot of the time people just do what everybody else is doing and accept society's rules and values. Boyd and Richerson (1985) constructed theoretical models to investigate the evolution of this conformity and found that virtually all of the circumstances that favour a reliance on social learning will also lead to very strong conformity. This

---

[6] This work builds on an earlier analysis by Cavalli-Sforza and Feldman (1973).

theoretical finding is supported by considerable empirical evidence. Among animals and human beings alike, when individuals learn from others they frequently tend to do what the majority of the population are doing (Boyd and Richerson, 1985; Laland *et al.*, 1996b). One consequence of conformity is that it makes it hard for new behaviour to spread within a population, as only common variants are favoured by cultural selection. This means that if groups of individuals differ in their learned behaviour, conformity will act to maintain these differences while at the same time minimizing differences in behaviour within groups.

For Boyd and Richerson (1985), group selection operates on culture. Thus, it is not genes that are selected for but rather groups of individuals expressing a particular culturally learned idea or behaviour. To give an example, imagine a population of individuals that cooperate to build a stockade to protect themselves in times of conflict. If this population suffers many fewer wartime casualties than other groups that don't build stockades then its numbers may increase to the point where it gives rise to new populations at a faster rate than enemy populations. Provided these daughter communities also build stockades then this cooperative behaviour may become widespread across broad populations. There are no genes for building stockades involved: rather, group selection has favoured the culturally transmitted idea of the stockade.

Several properties of cultural transmission, as opposed to genetic inheritance, make Boyd and Richerson's idea attractive. First, conformity helps maintain group differences. Group selection of group-beneficial cultural preferences and knowledge is possible because the transmission process discriminates against non-conformers. For instance, a tendency to do what the majority are doing compels indivi-

GENE–CULTURE COEVOLUTION 265

duals to cooperate to build the stockade. Secondly, at the group level, selection of cultural variants can be faster than selection of genetic variants because a threatened or defeated people may adopt the cultural knowledge and preferences of a new conquering culture, either voluntarily or under duress. Thus, unlike the group selection of genes, here inclusion of new individuals in the group will not necessarily weaken the process. Thirdly, symbolic group marker systems, such as human languages, cultural icons, totems, and flags, make it considerably easier for cultures to maintain their characteristic features and to resist imported cultural information from immigrants than it is for local gene pools, known as demes, to maintain their genetic differences by resisting gene flow. Fourthly, cultural transmission of information about cheaters, such as gossip, together with socially sanctioned forms of punishment for cheating, removes the advantages of non-cooperation. The net result is an increase in the strength of group selection.

Whether Boyd and Richerson's hypothesis can work depends on rates of group formation and group extinction. Soltis *et al.* (1995) put the theory to the test by using data on these rates among small communities in New Guinea. Their analysis led to the conclusion that, if the measured extinction rates were representative, cultural group selection was potentially a good explanation for slowly changing aspects of culture such as social structure, conventions, and institutions, but not for more rapidly changing fads.

However, group selection may have a more disturbing side. In truth, group selection does not directly favour altruistic individuals so much as 'selfish' groups. Selection between cultural groups may engender hostility and aggression to members of other groups, fear of strangers, slanderous propaganda concerning outsiders, and so on. The same process may simultaneously be responsible for both the best of human motives and the worst attributes of human societies. Richerson and Boyd (1998) argue that a long history of cultural group selection would have created the social environment that favoured the selection of genetic predispositions for altruistic behaviour to in-group members and also hostility to outsiders, which they label 'tribal instincts'. Their analysis demonstrates that, when cultural transmission is included into evolutionary models, the nature of the evolutionary process may be quite dramatically different.

Heritability of intelligence and personality traits

Scientists have tried to unravel to what degree differences among humans in intelligence, cognition, and personality traits are due to genetic factors and to what degree they reflect other influences, such as the developmental environment, learning, or culture. The extent to which differences in a trait are the result of genetic differences between individuals is commonly encapsulated in a measure known as *heritability*.[7] As an example, consider a characteristic such as body weight, which may depend on genetic factors but also on nutrition, exercise, and other environmental influences. If we consider a particular population of human

---

[7] Geneticists distinguish between two uses of the term heritability (Falconer and Mackay, 1996). A broader use relates to that proportion of the variance of a phenotypic trait that is caused by genetic factors (V_G/V_P). A narrower usage, which specifies the fraction of phenotypic variance that can be attributed to variation in the additive effects of genes, relates exclusively to those genetic influences that are transmitted from the parents and affect the response to selection (V_A/V_P). It is this narrower usage that is employed in most behaviour genetics analyses and to which we refer here.

beings, we can ask how much of the variance in their weight is down to variation in people's genes? The proportion of the total variance attributed to genetic effects is the *heritability ratio*, which can vary from 0 to 1. If everyone in the population had the same genes, and all differences in weight could be put down to diet and other environmental effects, heritability would be 0. At the other extreme, if people differ in weight solely because they have different genes, then the heritability ratio would be 1. Thus heritability is not a measure of the importance of genes in the development of a character, but rather a measure of to what extent alternative genes explain the differences between people.

If we were to measure the heritability of weight in a sample of people who were all well fed and who experienced a similar upbringing, the heritability might be quite high as there would have been few environmental differences to cause variation in weight. Within another population, which had experienced greater variation in diet and environment, we might measure a considerably lower heritability for weight. Thus the heritability of any given characteristic is not a fixed and absolute quality but a property of the population sampled. Moreover, characteristics can run in families, but have low heritabilities. For instance, this is the case for musical ability, where the differences between individuals are more down to practise, a supportive family environment, and good teaching than genes (Bateson and Martin, 1999). Unfortunately, attempts to estimate heritabilities are frustrated by the fact that little is known about how genes and environment combine to shape a developing individual (Bateson and Martin, 1999). As a result, estimates of heritability are also highly dependent on the formal model used to compute them (Feldman and Otto, 1997).

Identical twins have exactly the same genotype so it

might be thought that the frequent similarity of their appearance and behaviour reflects their common genes. However, any resemblance is also partly caused by the twins experiencing a similar environment. Consequently, most estimates of heritability for behavioural traits in humans are based on studies that compare genetically identical (monozygotic) twins with fraternal (non-identical or dizygotic) twins (Bouchard *et al.*, 1990; Plomin *et al.*, 1993). Researchers who estimate heritabilities, known as 'behaviour geneticists', commonly begin by assuming that the degree of similarity of environment experienced by a pair of identical twins is on average roughly the same as the degree of similarity of environment experienced by a pair of fraternal twins. They then go on to propose that any greater degree of resemblance in the traits exhibited by identical twins compared to fraternal twins must reflect the greater genetic similarity of the identical twins. However, the assumption that identical and fraternal twins experience equivalent amounts of shared environments is questionable. Identical twins may be treated more similarly by others than are fraternal twins. Identical and fraternal twins differ in the extent to which they share similar environments in the womb. Moreover, identical twins will always have a same-sex twin, whereas a fraternal twin could be brother or a sister, which may result in a very different relationship between the siblings. Heritability studies based on twins alone do not provide sufficient data to disentangle genetic from cultural influences and, as a consequence, estimates of heritability based on twin studies are generally inflated (Feldman and Otto, 1997; Devlin *et al.*, 1997). Researchers frequently make other simplifying assumptions in twin studies, such as that there are no interactions between genes (epistasis) and that there are no gene–

environment interactions. It is also commonly assumed that identical twins raised apart because of adoption share no environmental or cultural similarities. In reality, adoptive placement is far from random and often occurs after an extensive period in which the children are together (Goldberger, 1978).

Gene–culture methods have been put to use to make sense of this tangled issue. Following the early work by Cavalli-Sforza and Feldman (1973), Sally Otto, an evolutionary biologist at the University of British Columbia, Vancouver, together with colleagues Freddie Christiansen and Marc Feldman (1995) combined gene–culture models with other statistical methods to investigate how genetic and cultural effects on behaviour are transmitted between generations. They considered the effects of a variety of different mechanisms of cultural inheritance and also included the possibility of individuals being biased in their choice of sexual partner, as well as the influence of non-transmitted environmental factors. Their models were applied to complex data on the transmission of personality traits within families and have provided some of the most sophisticated analyses of heritability to date.

An important point that emerged from this work is that heritability estimates are extremely sensitive to the assumptions that are contained in the model. For instance, Otto *et al.* (1995) estimated the values of the parameters in their models using data on correlations of IQ (a measure of intelligence) within families. This is a means of testing the assumptions of the behaviour geneticists. For instance, if identical twins really do experience an equivalent degree of similarity in their environments to fraternal twins, then any parameter that represents the difference between the levels of similarity for the two types of twin will be estimated to be close to zero. However, this is not what they found. Using data from 111 studies of IQ which were collected together by Bouchard and McGue (1981), they found that, for the models to give a good fit to all of the data, they had to include a parameter that represented the degree to which identical twins experience a more similar environment than fraternal twins. As part of environmental experience is how one is treated by other people, perhaps identical twins are treated more similarly by others than twins who look different from each other[8] (Feldman and Otto, 1997). The finding questions how valid and widely applicable are heritability estimates based solely on twin-study data. When it comes to estimating heritability, twins are an unrepresentative source of data about the entire population, most of whom are not twins, and genetically identical individuals are even more unrepresentative.

In general, the model which gave best fit to the IQ data included a large influence of shared environment parameters. Ignoring these parameters, as is common in the behaviour genetics literature, leads to a significant drop in the goodness of fit of the model to the data, a marked increase in heritability estimates, and a corresponding reduction in the variance attributed to culture. Otto *et al.* (1995) obtained heritability estimates for IQ of around 0.3, which mean that only 30% of the variance in IQ can be put down to genetic differences between the people sampled. This contrasts starkly with the inflated estimates generated using

---

[8] Another possible explanation is the existence of genetic interactions (epistasis) that make genetically identical individuals more alike, but which contribute little to the resemblance of other relatives (Feldman and Otto, 1997).

twin data alone, which range from 0.6 to 0.8, and would erroneously suggest that most of the variance in IQ is due to genetic differences between people. A similar picture emerged from Otto *et al.*'s (1995) analysis of other personality traits. Once again, heritability estimates relying exclusively on twin or parent–offspring correlations were found to be inflated. Otto *et al.*'s findings throw considerable doubt on the claim that personality variables are not influenced by social learning (Eaves *et al.*, 1989).

Critical evaluation

With the exception of Lumsden and Wilson's work, gene–culture coevolutionary theory has not been subject to the same level of criticism as other evolutionary approaches. In fact, it has been almost completely ignored in the debates over human sociobiology and her progeny, perhaps because of its technical nature. However, some social scientists have objected to the idea that culture can be modelled as if composed of discrete psychological or behavioural characteristics, while others have questioned the legitimacy of 'borrowing' population genetics processes to model culture. Additionally, researchers from many backgrounds have suggested that biological evolution is too slow and cultural change too capricious for their interaction to be genuinely coevolutionary. Finally, while gene–culture coevolution has a strong and rigorous theoretical foundation, it is vulnerable, as is memetics, to the charge that it has not spawned an empirical science. The principal problem for gene–culture coevolution is that, to date, only a handful of mathematically minded scientists around the world are actually doing it. In this section, we discuss these criticisms in turn.

Can 'culture' be subdivided into discrete units?

In 1871, Edward Tylor, the leading anthropologist of his day, defined culture as:

> that complex whole which includes knowledge, belief, art, morals, custom and any other capabilities and habits acquired by man as a member of society.

Tylor's rather cumbersome definition was to hold sway over the anthropological community for many decades and even today it captures the intuitive notion of culture held by the lay person. However, if culture is an amorphous, interwoven conglomerate of knowledge, behaviour, and tradition, it is difficult to envisage how it can be modelled as if it were transmitted between individuals in simple, clean packages or as transformed distributions. Clearly there has been no attempt on the part of gene–culture coevolution to track the entire culture of a people. Nor is the goal of gene–culture coevolution to model stages of societal progression or complexity, a historical focus of anthropology and 'cultural evolution' approaches (see Chapter 2). Fortunately, onerous and all-encompassing definitions of culture, like Tylor's, have had their day. More cognitive perspectives are in the ascendancy which restrict culture to learned *information* stored in the brain and transmitted between individuals (Durham, 1991; Goodenough, 1999). As the 'culture' in gene–culture coevolution is socially learned information, the cognitive revolution would seem to have paved the way for studying and quantifying culture in a manner similar to that employed in gene–culture analyses.

However, there is a catch, which remains a major stumbling block to many social scientists who might otherwise embrace the methods of gene–culture coevolution. Can culture really be chopped up into units? For the majority of social scientists, it is too simplistic to analyse people's behaviour or ideas one at a time as there are too many other interacting factors (Bloch, 2000). The current fashion lays emphasis on a more qualitative, holistic description of cultural phenomena and is very suspicious of formal models. Can the culture of a people be treated like a collection of beans in a bag?

In truth, neither the 'complex whole' nor the 'beanbag' representation is a truly accurate description of the system. In fact, exactly the same debate has taken place within evolutionary biology, with Mayr (1963) criticizing the 'beanbag genetics' assumptions of theoretical models and Haldane (1964) responding with a vigorous defence. As with culture, gene interaction during development is neither amorphous nor inextricably interwoven, yet most population geneticists are in no doubt that Haldane's version has proven useful (Crow, 2001).

We have little sympathy with the obscurant holism that afflicts many of the social sciences. To use another analogy, the human brain is also a complex and interconnected system of interacting processes. Yet this has proved no barrier to the unstinting march of neuroscience, which has made phenomenal progress in understanding brain functioning, often by employing extremely crude methods such as brain lesions or injection of neurotransmitter-blocking drugs. The intricacies of the big picture are laid bare one small step at a time. Gene–culture researchers recognize that culture is an elaborate and diverse entity. Yet the fundamental lesson of science is that patient chipping away at such perplexingly intricate problems yields dividends in the long run.

In each gene–culture study, individuals have been classified according to whether they possess a particular package
of psychological constructs; for example, whether they
believe dairy products are good to eat, know sign language,
or prefer sons to daughters. This is conceptually no different from focusing on a particular gene and classifying individuals according to genotype. It does not mean that all
other aspects of an individual's culture are irrelevant, but
rather that it is instructive to consider the average effect of
the particular information across the entire population. For
gene–culture enthusiasts, breaking down culture into units
is merely a useful theoretical expedient. Perhaps at this
formative stage researchers should be content to concentrate on those dynamic cultural phenomena that are well
described by changes in the frequencies of packages of
information and to leave to one side, for the moment, cultural phenomena that are too messy to be depicted that way.
It may be that some areas of culture are more easily
chopped up than others and hence more amenable to this
kind of analysis. There is no shortage of poorly understood
cultural entities that would benefit from quantitative
investigation.

As with the piecemeal approach of human behavioural
ecologists, the bottom line is that biologists and human
scientists alike will not be able to understand cultural
processes unless they are prepared to break them down into
conceptually and analytically manageable units. A glance
through any undergraduate textbook on psychology reveals
that there is considerable and compelling evidence that
humans acquire packages of learned and socially transmitted information, store them as discrete units, chunk and
aggregate them into higher order knowledge structures,
encode them as memory traces in interwoven complexes of

neural tissue, and express them in behaviour.[9] It is not such an extraordinary claim that culture is acquired in bits and pieces.

How similar are genetic and cultural processes?

Social scientists and biologists alike have noted analogies between the processes of biological evolution and cultural change, the gene and the symbol, the gene pool and the 'ideas pool' (Campbell, 1974; Cavalli-Sforza and Feldman, 1981; Plotkin and Odling-Smee, 1981; Boyd and Richerson, 1985; Durham, 1991; Goodenough, 1999). For instance, both genes and memes are informational entities that are differentially transmitted as coherent functional units and that exert an influence on the phenotype. Naturally, there are also differences between these phenomena, so it would not be appropriate simply to take theories from one field and apply them naively to the other. However, the analogies have proven sufficiently tempting and the interactions sufficiently rich to prompt the development of conceptual and formal dual- or multi-level inheritance models. Borrowing Darwinian concepts and methods, suitably adjusted to the structural peculiarities of human culture, is the quickest and easiest path to a reasonable theory of human culture and thus to an improved understanding of human behaviour (Boyd and Richerson, 1985).

---

[9] Compelling evidence for this assertion can be found in the category-specific naming impairments of human patients with brain damage, currently subject to considerable interest within cognitive neuropsychology. Patients have been reported to recognize and correctly name all items except those in a specific category, such as fruit and vegetables, country names, animals, or objects related to the hospital environment (Crosson et al., 1997). Such studies suggest that at least some learned knowledge stored in human brains is organized into discrete semantic categories.

While it may be of great and legitimate concern to others (Plotkin, 1994; 1997), we regard the debate over whether or not the analogies represent an underlying similarity of process as a red herring. Dual-level models could be constructed even if there was no resemblance at all between the two levels. In fact, much of what makes culture interesting derives from its differences from genetic inheritance. The books by Cavalli-Sforza and Feldman (1981) and Boyd and Richerson (1985) are primarily concerned with how the dynamics of cultural evolution differ from that of genes. Ultimately what matters is whether the models that have been constructed are good models in the sense that they capture the essential properties of the system. If either cultural or biological processes do not operate in the way assumed by the models, new models could be developed with assumptions that can be more readily justified. In contrast, approaches that focus on a single process (be it exclusively cultural or exclusively genetic) have made the fundamental and sweeping assumption that there is only one process that matters, or that the processes do not interact.

Researchers from both the human behavioural ecology and evolutionary psychology schools have criticized gene–culture coevolutionary theory as promoting a false dichotomy of culture being distinct from biology (Flinn and Alexander, 1982; Daly, 1982; Tooby and Cosmides, 1989; Flinn, 1997). They argue that as the capabilities for cultural acquisition and retention are evolved, then cultural processes must be adaptive and cannot be de-coupled from biology. We suspect that there are few real differences of opinion here between gene–culture advocates and critics. Certainly those gene–culture researchers to whom we have spoken do not believe that culture is an entirely self-determined process, isolated from biology, although the criticism could perhaps be made to stick on some advocates of memetics. On the contrary, gene–culture models are devised precisely to explore how biological predispositions shape cultural learning and how cultural processes modify selection pressures on genes. The models do not attempt to separate nature from nurture but to simplify the system to a sufficiently tractable level to be easily understood. If there is a difference of opinion it relates to whether the single process of biological evolution is sufficient to account for cultural variation or whether a second process of cultural transmission is also necessary. Several gene–culture co-evolutionary analyses have provided evidence that single process models do not explain data as well as do gene–culture models (Otto *et al.*, 1995; Laland *et al.*, 1995a, b), that equivalent single process models either have (or would have) reached erroneous conclusions (Kumm *et al.*, 1994; Feldman and Cavalli-Sforza, 1989), and that the interaction between genes and culture can change the evolutionary process, for instance by generating a new form of group selection (Boyd and Richerson, 1985) or by modifying evolutionary rates (Feldman and Laland, 1996). These are compelling reasons to treat transmitted culture as a potent process in the shaping of human evolution.

Do genes and culture coevolve?

It is frequently suggested that genetic evolution is too slow, and cultural change too fast, for the latter to drive the former (e.g. Adenzato, 2000). In fact, as we noted in the evolutionary psychology chapter, artificial selection experiments and estimates of the strength of natural selection in plant

and animal populations that are currently evolving reveal
that biological evolution may be extremely fast. Significant
genetic and phenotypic change is sometimes observed in a
small number of generations. At the same time, observations of hominid stone tool technologies reveal that cultural
change can be extraordinarily slow. Acheulian and
Oldowan stone tool traditions remained very similar for
hundreds of thousands, even millions, of years. Even cultural institutions such as labour markets can be extremely
persistent, albeit on a shorter time scale (Bowles, 2000).
Furthermore, theoretical analyses have revealed that cultural transmission may change selection pressures to generate unusually fast genetic responses to selection in humans
(Feldman and Laland, 1996). It is thus entirely feasible that
genetic and cultural evolution could sometimes operate at
similar rates. In fact, the past two million years of human
evolution may even have been dominated by gene–culture
coevolution.

Durham (1991) illustrates with compelling examples, each
backed by considerable data, how variability in human behaviour and society may be interpreted as resulting from interactions between genetic and cultural processes. Durham identifies five categories of interaction: (1) *genetic mediation*,
where genetic differences underlie cultural variation, as may
be the case for the terms used by humans to describe colour,
which reflect features of the human visual system; (2) *cultural
mediation*, where culture drives genetic change, such as with
the evolution of adult lactose absorption in populations that
consume dairy products; (3) *enhancement*, where culture
reinforces genetic predispositions, as with the emergence of
incest taboos that guard against the deleterious effects of
inbreeding; (4) *neutrality*, where cultural variants are adopted
independently of an individual's genotype, as is the case for

learning different languages;[10] and (5) *opposition*, where culture leads to maladaptive traditions, for instance the cannibalism of the Fore, a New Guinea community, that spread the deadly nerve disease *kuru*.

Culture can, of course, cause rates of environmental change that really are too fast for human genetic evolution to track, and it is probably doing so increasingly. In fact, in the last twenty-five to forty thousand years the dominant mode of human evolution has probably been exclusively cultural. However, that does not mean there has been no evolutionary feedback from culture: it merely switches the evolutionary response to the cultural domain. Cultural niche construction favours further cultural change, perhaps at accelerating rates, with coevolution occurring between culturally transmitted characters (Odling-Smee *et al.*, 2000).

Could there be an empirical science of gene–culture coevolution?

In the previous chapter we posed the question 'Is a science of memetics possible?' A little reflection revealed a considerable array of empirical methods that could be applied to study memetics, were meme enthusiasts so inclined. The trouble is, few people are actually engaged in the business of counting, recording, and measuring cultural variants or in

---

[10] Anthropologist Edmund Leach (1981) regarded human language as the biggest obstacle to Lumsden and Wilson's gene–culture coevolution. As any child could learn the language of another culture there would not seem to be any evolved predispositions biasing the adoption of specific cultural variants. Durham's scheme, which is more representative of modern gene–culture coevolution than Lumsden and Wilson's, allows for a much broader interpretation of the relation between genetic and cultural variation, and includes such cases of neutrality.

tracking how they change in frequency. It would seem that the same problem bedevils gene–culture coevolution, which is destined to remain an under-utilized branch of research into human behaviour and evolution unless it generates a programme of empirical science.

Currently the methods of gene–culture coevolution are almost entirely theoretical in nature. A psychologist who was inclined to adopt Tooby and Cosmides's (1989) formulaic procedures for doing evolutionary psychology could potentially do so tomorrow. However, gene–culture coevolution provides opportunities for young researchers to develop their own models if they are willing to gain the relevant background in, say, anthropology or psychology, and to become sufficiently familiar with theoretical population genetics. They can then help to bring the gene–culture coevolutionary approach to areas as yet unexplored by the current researchers. At the moment, the gene–culture coevolutionary approach is largely confined to Marc Feldman and Luca Cavalli-Sforza, Rob Boyd and Peter Richerson, and a small number of other researchers, including Ken Aoki, Bill Durham, Jochen Kumm, Kevin Laland, Sally Otto, Alan Rogers, and Lev Zhivotovsky.

Are there no clear empirical predictions of gene–culture coevolution? In fact, where gene–culture analyses have been applied to specific case studies they do make a variety of testable predictions. For example, Aoki *et al.* (1996) detail the conditions under which the spread of farming will generate predictable geographical patterns in gene frequencies. Laland's (1994) model of sexual selection with culturally transmitted preferences implies that there should be society-specific correlations between anatomical traits in one sex and learned preferences for the traits in the opposite sex. Henrich and Gil-White (2001) explore the evolution of

prestige, in the process making a valuable empirical contribution to gene–culture coevolution. Other empirical predictions and findings can be found in Feldman and Laland
(1996). Moreover, as our discussion of Otto *et al.*'s (1995)
work illustrated, gene–culture coevolutionary methods can
also be used to unravel patterns of inheritance in behavioural and personality traits. There is much to be gained
from empirical researchers interested in human evolution
or human behaviour genetics employing established
gene–culture coevolutionary models to interpret their own
data.

However, as yet there is no well established general
empirical method for doing gene–culture coevolution. The
closest to such a general approach is that advocated by Bill
Durham, a researcher who has done more than most to
pioneer an empirical science of gene–culture coevolution.
Durham (1991) argues that the main, but not exclusive,
means of cultural change is cultural selection guided by prior
cultural knowledge and beliefs. This leads to the prediction
that knowledge of prior beliefs will be both necessary for
explaining the direction and rates of cultural change and
sufficient to explain why some memes spread rapidly while
others peter out. Durham also predicts that the memes of
highest cultural fitness will also tend to be those of highest
inclusive fitness for their selectors. This contrasts with the
view of most meme enthusiasts that memes spread entirely
independently of their Darwinian fitness. While Durham's
hypotheses are not shared by all researchers advocating the
gene–culture coevolutionary approach, they illustrate how
general coevolutionary predictions can be devised and put to
the test.

So there is considerable potential for an empirical science
of gene–culture coevolution. For instance, many of the

experimental methods mentioned in the meme chapter would also be of benefit here, including studies to determine whether there are correlations between classes of cultural information, and whether there is cultural character displacement and convergent conceptual evolution (see Chapter 6 for details). We will end with two further examples of investigations that illustrate how the findings of the gene–culture research programme can be put to the test.

The first involves the carrying out of experiments, using transmission chain procedures, which explore to what extent and in which manner cultural information evolves in a group of changing composition (e.g. Jacobs and Campbell, 1961; Insko *et al.*, 1982; 1983). For instance, Insko *et al.* (1983) studied the cultural evolution of inter-group relations in miniature societies created in the laboratory. They established groups of four people, removing and adding one person per 'generation' to mimic death and birth processes. In one experiment subjects were given a sham IQ test, and then divided into 'smart' and 'dumb' groups ostensibly on the basis of their IQ scores but actually at random. The smart groups were given control of subject payment to mimic the conquest of one group by another. Over the course of the experiment the dominant 'smart' groups not only began to treat the 'dumb' groups unfairly but evolved rationalizations for their actions, while subordinate groups evolved counter-strategies such as going on strike or deliberately slowing down their work. Insko's study not only provides laboratory evidence for cultural evolution but is strikingly consistent with Richerson and Boyd's (2001) tribal instincts hypothesis.

Support for the idea of psychological adaptations to group living can also be found in an exciting cross-cultural study of experimental economics carried out by a large team of anthropologists and economists known as the *MacArthur Foundation Research Group on the Nature and Origin of Norms and Preferences*, directed by Rob Boyd and Herbert Gintis. These researchers recruited subjects from fifteen small-scale societies in twelve countries around the world and paid them to play economics games (Henrich *et al.*, 2001). In one of these, the ultimatum game, a subject is provisionally assigned a substantive sum of money (equivalent to a day or two's wages) to share with another individual. The 'proposer' is allowed to make a single offer of a proportion of the stake to a 'respondent', who has one chance to accept or reject it. Acceptance means each participant keeps his or her share, while rejection results in neither party receiving anything. The most rational behaviour would be for the proposer to offer the bare minimum amount (as this maximizes the proposer's share) and for the respondent to accept it (since something is better than nothing and there is no chance for negotiation). However, the ultimatum game has been subject to extensive study by economists (generally using university students), who find that proposers consistently offer more than the bare minimum while respondents commonly reject the proposer's offer. Students make predictable and relatively invariant offers, consistent with income maximization given the assumption that low offers will be rejected, with the most common (modal) offer being 50%. In contrast, the MacArthur Foundation study found considerable variation in the mean offer between societies, ranging from 26% among the Machiguenga of Peru to 58% among the Indonesian Lamelara, with sample modes varying from 15 to 50% (Henrich *et al.*, 2001). Rates of rejection were also much more variable than previously observed among students.

The large variation across the different cultural groups strongly suggests that preferences and expectations are affected by group-specific conditions, such as social institutions or cultural norms concerning fairness. For instance, until recently, Machiguenga families were almost entirely economically independent of each other and rarely engaged in productive activities outside the family. In contrast, the Lamelara are cooperative whale hunters who go to sea in large canoes manned by a dozen or more people. A plausible interpretation of the subjects' behaviour is that, when faced with a novel situation (the experiment), they looked for analogues in their daily experience and then acted in a way appropriate for the analogous situation. Generous offers were found in societies with a culture of gift-giving while stingy offers were found among peoples not used to sharing.

How are these findings to be interpreted? Evolutionary psychologists would surely have anticipated that people the world over would have behaved more uniformly on the assumption that they all share the same evolved psychological mechanisms. Human behavioural ecologists would almost certainly have started with the 'rational actor' model that anticipates minimal offers and rejections, and only subsequently converged on more realistic models. Researchers working within the gene–culture tradition, on the other hand, would start with the assumption that there will be society-specific norms that influence performance in tasks such as these games and which vary across societies. It is only when the cultural traditions of the population are taken into account that the behaviour of subjects can be satisfactorily interpreted. Norms and social institutions typically change comparatively slowly (Bowles, 2000), indeed slow enough to be within the range of phenomena that can be explained by cultural group selection (Soltis et al., 1995). It remains an intriguing possibility that the observed variation in society-specific norms of fairness and social institutions is the product of cultural group selection (Boyd and Richerson, 1985).

Conclusions

Gene–culture coevolutionary analyses suggest that evolution in species with a dynamic, socially transmitted culture may be different from evolution in other species, for at least three reasons. First, culture is a particularly effective means of modifying natural selection pressures and driving the population's biological evolution, as was the case for lactose absorption. Secondly, culture may generate new evolutionary processes, for instance cultural group selection. Thirdly, cultural transmission may strongly affect evolutionary rates, sometimes speeding them up and sometimes slowing them down. Such findings suggest that traditional evolutionary approaches to the study of human behaviour may not always be adequate. As we have seen, there is considerable potential for further empirical work on gene–culture coevolution. Even the mathematical methods involved can be accessible to non-mathematicians.

Certainly, there are challenges that currently limit the application of gene–culture methods. For instance, the models assume a correspondence between the socially learned information that individuals acquire and their behaviour, yet people's actions are not always consistent with their beliefs (Cronk, 1995). In addition, gene–culture methods have paid comparatively little attention to exploration of interactions within families (such as nepotism and parent–offspring conflict), which have been so fruitful for other evolutionary perspectives. These drawbacks are not insurmountable but no-one has yet devised methods for dealing with them. Nonetheless, we are not alone in seeing rich possibilities for the utilization of gene–culture coevolutionary methods by biologists and social scientists. We leave the last word to Edward Wilson:

> It is possible that gene–culture coevolution will lie dormant as a subject for many more years, awaiting the slow accretion of knowledge persuasive enough to attract scholars. I remain in any case convinced that its true nature is the central problem of the social sciences, and moreover one of the great unexplored domains of science generally; and I do not doubt for an instant that its time will come. (1994, p. 353)

CHAPTER 8

Comparing and integrating approaches

We can now return to the question of whether evolutionary theory can help us to understand human behaviour and society. We have seen that the history of taking an evolutionary perspective on human behaviour has been filled with misdemeanours that cannot be ignored. Yet there is also evidence that the careful use of evolutionary theory can increase our understanding of humanity. The preceding chapters have established that there are numerous ways to exploit evolutionary theory to investigate human behaviour, each of which has provided valuable and novel insights. While it is common for each school of thought to portray its methods and reasoning as the way and the light, in truth each approach has its strengths and weaknesses. But how do they all fit together?

Our aim in this final chapter is to compare the five approaches, to discuss to what extent they are complementary, and to explore what part each can contribute to the complete picture. However, having delineated research into human behaviour and evolution into particular schools, it is only appropriate that we stress that the field is not quite as easy to partition in real life. For instance, our portrayal of human sociobiology is to a large extent a historical account, and over the last three decades this discipline can be regarded as having influenced the formation of, or even dissolved into, the subdisciplines of human behavioural ecology, evolutionary psychology, and gene–culture coevolution (Smith, 2000). Even among the four contemporary approaches there is much common ground and considerable overlap in perspective and methodology. We accept that the exercise in which we are engaged, which attempts to crystallize the distinct clusters of view, will inevitably create the impression that the boundaries between these schools of thought are cleaner than they actually are. Without doubt, the reality is a good deal messier, and the various disciplines share considerable common ground (Daly and Wilson, 2000). Far from wishing to establish or reinforce artificial frontiers, we would like to encourage the evolutionary minded to move selectively between schools, picking and choosing the best tools available, drawing insights from each and synthesizing divergent perspectives in a critical and discerning manner. To this end, we believe evolutionary enthusiasts are more likely to be successful if they are aware of the merits of each school, and of the pros and cons of each method.

This chapter begins with a discussion of the extent to which each of these approaches is currently being used by researchers. Focusing on the example of infanticide, we show how a single topic can be investigated separately from each of the five perspectives. We then discuss how simultaneous use of the different approaches may provide the broadest explanation of human behaviour, using the study of war as an example. Finally, in the concluding section of the chapter, we explore whether the schools are complementary, or whether there are some fundamental differences of opinion that prevent their coming together.

Which is the most popular approach?

Let us indulge in some playful 'meme's-eye view' reasoning, and ask 'Which of these five evolutionary approaches is the most infectious meme?' Which school of thought is winning the battle to monopolize the hearts and minds of evolutionary orientated human scientists? Of course, popularity, in terms of the number of researchers or published scientific papers, does not necessarily provide a measure of which research programme is best. However, the schools that we have described emerged almost simultaneously and hence one could argue that researchers have voted with their feet as to which methods are judged the most useful or exciting.

The first obvious conclusion is that it is not the 'meme' meme that is winning the popularity stakes, at least not currently. However prevalent memetics may be amongst the casual readers of popular science or in the bulletin board discussions of internet users, it has yet to make a serious impact on the sciences or social sciences (Aunger, 2000). Academics almost appear to have been inoculated against the meme virus, perhaps because they find meme's-eye view reasoning disturbing or can't see how to turn it into an experiment, or perhaps because they find biological accounts of human behaviour more attractive.

Neither, despite its high profile beginnings, is it human sociobiology, which, justified or not, has accumulated an unwelcome and sinister baggage of tightly linked deleterious memes such as 'genetic determinism' and 'prejudice'. Currently, many biologists appear wary of describing themselves as human sociobiologists for fear of attracting hostile criticism, while among social scientists sociobiology is frequently subject to a silent loathing. For many, the best that many would-be sociobiologists can do is go underground, adopt a pseudonym like 'evolutionary psychologist' or 'human behavioural ecologist', and hope that the passage of time will one day allow their views to be judged more objectively.

Least catchy of all memes is the mathematical world of gene–culture coevolution. In many ways the most complex and potentially rewarding of all approaches, this package, with its multiple processes and cerebral onslaught of sigmas and deltas, may appear too abstract to all but the most enthusiastic reader. Until such a time as the theoretical hieroglyphics can be translated into a respectable empirical science most observers will remain immune to its message. However, such empirical methods are available and, as we saw in Chapter 7, are starting to have an impact.

The importance of a vibrant empirical programme might be the primary lesson that gene–culture enthusiasts could learn from human behavioural ecology, whose star, in spite of its theoretical bent, has risen to a respectable position in the heavens. While not exactly an exemplar of memetic catchiness, human behavioural ecology has none the less spawned a healthy empirical industry, perhaps partly by sidestepping the potentially contentious issue of the genetic bases of human behaviour.

Undoubtedly the reigning champion of Darwinian memes is evolutionary psychology, with the Santa Barbara virus seemingly the most infectious of all known strains. Perhaps its success can, in part, be attributed to the ease with which this perspective can be translated into scientific research, which in turn renders it highly visible and more readily adopted. Arguably, the focus on human universals resonates less threateningly with an audience quick to associate evolutionary explanations for human differences with racism and genetic reductionism. Without doubt, evolutionary psychology has been blessed with some extremely talented writers (notably Steven Pinker and Robert Wright), and many evolutionary psychologists are very good at producing semipopular accounts that receive much media attention and may attract unaligned academics. Moreover, evolutionary psychology provides clear answers to questions concerning contemporary Western societies. To whatever we attribute its prosperity, evolutionary psychology is undoubtedly the dominant school of thought.

An example of complementary information

The five different approaches are well illustrated by focusing on a single topic that all have independently addressed. This exercise highlights the complementary nature of the information generated by these different methods, each contributing to a broad general understanding. One such useful example is the topic of human infanticide, more specifically, infanticide by mothers or by unrelated males.

In Chapter 3, we described how Sarah Blaffer Hrdy, working within the sociobiological tradition, was able to make sense of the curious observation that female monkeys will mate with infanticidal males as an adaptive strategy on the part of females in response to the high turnover of males in the group. Hrdy (1999) argued that infanticide by mothers is more common in human beings than is infanticide by 'invading' males. As human infants are very costly to raise, in terms of time, energy, and resources, mothers require the cooperative help of social companions, as is the case in some species of New World primates. The ability of a mother successfully to raise a child may therefore be contingent on the amount of social support that she receives.

Here, a comparative perspective which considers human infanticide in the context of that exhibited by other primates has helped shed light on those aspects of this behaviour that are similar to, and different from, those of closely related species. Hrdy (1999) also discusses how many of the traits that make babies and infant primates attractive to adults may have evolved in response to a past history of neglect or abandonment by parents. Such features that make infants attractive may be counter-strategies that have evolved because they reduce the risk of infanticide by mothers. This cross-species comparative perspective is underused by all of the other approaches.

Human behavioural ecologists, described in Chapter 4, have studied human infanticide and neglect of offspring as part of a broader interest in parental care (Voland, 1998). They have hypothesized that, in a number of circumstances, mothers may be selected to terminate investment in a young infant and suggest that infanticide can be best understood as a strategy used to allocate limited resources optimally in a manner that maximizes lifetime reproductive success. Natural selection may have favoured mothers, and sometimes fathers or close relatives, who decrease their investment or even kill offspring where the costs of raising the infant are expected to outweigh the benefits (Voland, 1998). This may occur where the infant is deformed or very ill, where the offspring is of a particular sex, or where the health of the mother will be compromised by attempting to raise the offspring. Records collated in Ditfurt, Germany, between 1655 and 1939 indicate that infant mortality was much higher among illegitimate children than legitimate children, and that death rates were particularly high when the mother went on to marry a man who was not the father of the child (Voland and Stephan, 2000). This infanticide is not due to stepfathers, as the deaths occurred prior to re-marriage, but instead suggests an adaptive strategy on the part of the mother. Unrelated members of the social group may also kill infants, particularly when a primary care giver is no longer around. For example, a study of the Ache of Paraguay revealed that 5% of children born were victims of infanticide during their first year of life, and all children who lost their mothers during their first year of life were killed, often being buried in their mother's grave (Hill and Hurtado, 1996). Human behavioural ecologists can thus be seen to have set out to understand patterns of infanticide by assuming that human beings exhibit adaptive strategies that link behaviour to the environment.

In Chapter 5, we described Daly and Wilson's (1988) evolutionary psychology analysis of infanticide, in which they documented the fact that there was a significantly elevated risk to children residing with step-parents. Drawing on a large number of findings from evolutionary theory, and with consideration of the selective environment inhabited by ancestors, Daly and Wilson were able to make a number of insightful predictions, including that substitute parents would care less for children than natural parents because they are unrelated. In contrast to the human behavioural ecologists, these authors have not stressed that infanticide in current times should be thought of as an adaptive behavioural strategy.

While a mother can be sure that a baby is hers no matter what it looks like, the father cannot. Consequently, Gaulin and Schegel (1980) have argued that it could be to a baby's advantage to look like the father to encourage parental investment and discourage infanticide. Daly and Wilson (1982) suggested that mothers, relatives, and friends are disposed to comment that babies look more like their fathers than their mothers as a mechanism to reassure the father that he is the true sire. This evolutionary psychology perspective suggests that human beings may possess a propensity to link physical features of the child with paternity, a psychological mechanism that may have a selective history without necessarily performing any current function in Western societies. In a short article in the evolutionary psychology tradition that received considerable attention, Christenfeld and Hill (1995) reported that people were able to match photographs of 1-year-old children to photos of their fathers but not of their mothers. This raises the possibility that there is a physical basis to attributions of paternal resemblance. However, if 1-year-old babies unambiguously resembled their fathers, the father would also be certain when a child was not his and might be tempted to withhold resources or even to kill the infant. A theoretical analysis by Mark Pagel (1997) has shown that there may even have been selection for the concealment of any physical similarity to the father when the domestic father is often not the biological father. Thus the extent to which children resemble their parents will depend on levels of certainty of fatherhood over evolutionary time scales.

We found no examples of research into infanticide in the memetics literature.[1] However, as we pointed out in Chapter 6, most meme enthusiasts have a similar perspective, if a less theoretical one, to researchers who develop the mathematical models of cultural evolution reviewed in Chapter 7

---

[1] There are memetic accounts of the conceptually related topic of abortion (e.g. Aaron Lynch, 1996). However, such discourse amounts to little more than speculation, and to our knowledge these ideas have not been subjected to any serious empirical or theoretical analysis.

(Cavalli-Sforza and Feldman, 1981). A study by Nan Li, Marc Feldman, and Shuzhuo Li at Stanford University's Morrison Institute for Population and Resource Studies, illustrates the kind of investigation that more quantitatively minded memeticists could undertake. Li *et al.* (2000) used cultural evolution theory to develop models that predict how the sex ratio at birth in China is likely to change in the face of fluctuations in fertility and a strong, culturally transmitted preference for sons. This son preference is often manifest in excess female mortality through infanticide of daughters, underinvestment in daughters or female-biased sex-selective abortion, which skews the sex ratio towards a glut of males. They conducted a survey to measure rates of cultural transmission across generations, equivalent to measures of the infectiousness of the meme for preferring sons, in the rural areas of two Chinese counties, chosen as low and high projections for the country at large. Plugging these estimates into their models, they were able to predict that across the whole of China in 2020 the sex ratio at birth would range between 1.1 and 1.34, that is, between 110 and 134 males for every 100 females, depending upon the strength of the transmission of son preference. Currently, the birth sex ratio in China is around 1.14 (Clarke, 2000). To the extent that phenotypic traits and preferences are similar to memes, this study serves as an illustration of how the fidelity of meme transmission can be estimated with empirical studies and how meme frequency changes can be modelled mathematically and put to use to make valuable demographic projections. The analysis suggests that reducing the distortion in the sex ratio will depend critically upon weakening the transmission of the preference for sons. An earlier study out of the gene–culture coevolution tradition by Kumm *et al.* (1994) had gone one stage further to predict how culturally transmitted preferences for sons would affect the selection of genes that distort the sex ratio.

Irrespective of the methodological differences among the practitioners, there is little that is conflicting or incompatible about these findings. In fact, each investigation reinforces the others, collectively building up a panoramic view of the topic at hand that spans genetic to sociocultural levels of analysis and transects distant continents. Here is an advertisement for pluralism in evolutionary perspective. There is no reason for researchers to restrict themselves to a single research technique when, by and large, the different methodologies are highly complementary.

An example of an integrated approach

One example of a broad evolutionary approach that draws from many schools is ethologist Robert Hinde's investigations of war and propaganda (Hinde, 1991, 1997; Hinde and Watson, 1995). Earlier sociobiological accounts of war were unsatisfactory, as they tended to treat war as merely large-scale violence, failing in the process to distinguish between different levels of social complexity. More accurately, war encompasses a degree of centralized organization, the collective mobilization of individuals with prescribed roles, the use of propaganda, the widespread recognition of seemingly distinct and important group differences, and socially sanctioned injury of out-group members (Hinde and Watson, 1995). Hinde views war as a complex institution that involves the behaviour of individuals, their relationships with others in their social group, and processes that operate at the level of the culture or society. He suggests that simple analogies across these different levels of social complexity cannot be used to extrapolate from explanations of individual aggressive behaviour to those of coordinated, modern warfare. In other words, war is characterized by processes beyond the aggressive motivation of individuals. For Hinde, a comprehensive treatment of war also needs to take account of the psychological and social consequences of group membership (Hinde, 1991). Members of a social group will identify with one another, will be dependent upon others in the group, and will share norms and values of that society. Pervasive cultural factors, such as perceived national characteristics and religions, may also increase the probability of conflicts (Hinde, 1997), and the economic and technological state of the society may facilitate the outbreak of collective violence and war.

According to Hinde's treatise, biological predispositions such as a fear of strangers, aggressiveness, and a tendency to distinguish in- and out-groups do not cause war. However, these predispositions do play an important role, as they are exploited, for instance in the propaganda of mobilizing and abusive leaders, in ways that lead to an image of the enemy as different or evil and which sanctify aggression against adversaries. An individual's view of war may be influenced by everyday background factors, such as books and films, which can create the impression that war is a natural way to solve conflict that promotes the prestige and status of those individuals involved. The institution of war may therefore be understood best by examining the complex interactions between behaviour at the level of the individual, the psychological processes involved in interactions between individuals, and the cultural processes operating within the society.

Hinde's analyses raise a number of questions and posit some potential answers, which we have supplemented with other evolutionary ideas and methods that have been portrayed in this book. Here are a few of them. (1) 'How have

the relevant biological predispositions evolved?' Several
answers have been proposed, including the ideas that they
are a side effect of kin selection or reciprocity, or a consequence of cultural group selection. (2) 'How can negative
and hostile attitudes towards the enemy spread so quickly,
sometimes seeming to turn peaceful neighbours into brutal
murderers overnight?' Perhaps this occurs through horizontal social transmission and other cultural evolution
processes in a climate catalysed by predispositions and
prior knowledge that leave individuals particularly recep-
tive to propaganda messages. (3) 'Why are individuals
prepared to die for their country or religion?' Perhaps they
are victims of coercion or manipulation, perhaps a history
of cultural group selection has reinforced self-sacrificing
attitudes or perhaps, in spite of the risks, soldiers have high
reproductive success.

In today's world of international terrorism, when mili-
tant extremists are able to inspire ordinary people to go to
war or to sacrifice themselves, with a resultant loss of thousands of lives and heightened tension between communities, it has become imperative that researchers comprehend
what it is that makes people behave in this manner. A
greater understanding of the interplay between an individual's behaviour, relationships between individuals, and
cultural influences may have current practical applications
for the avoidance or resolution of conflicts between groups.
This analysis illustrates the fact that, to develop a plausible
account of complex human behaviour and institutions, we
may well need models that incorporate a host of distinct
evolutionary processes that have influenced biological predispositions, psychological mechanisms and cultural selection processes, in addition to an understanding of the
processes of social and economic change.

Comparing schools of thought

Researchers can only draw from all schools to the extent that they are methodologically complementary and consistent. Could the divergent views that we have portrayed be integrated into a single unified field? Or are the views of human sociobiologists, evolutionary psychologists, human behavioural ecologists, memeticists, and advocates of gene–culture coevolution mutually incompatible such that they will never sit side by side in a single evolutionary framework? Let us take another look at the five evolutionary styles portrayed in the previous chapters, isolate their key differences and explore whether these are sufficient to prevent integration. Table 8.1 depicts the principal level of explanation, methods for hypothesis generation and testing, and comparative models employed by each school. It also shows to what extent the different styles anticipate that human behaviour will be adaptive, and gives a sketch of each school's conception of culture and of humanity.[2] In the following sections, we briefly discuss the four of these comparisons that appear most clearly to differentiate the alternative views and ask to what extent these differences are barriers to a conceptual and methodological integration. The most striking differences occur in the level of explanation, the methods of hypothesis generation and hypothesis testing, and conceptions of the nature and importance of culture.

---

[2] This table and the surrounding discussion builds on a similar comparative analysis by Eric Alden Smith (2000). See Borgerhoff Mulder *et al.* (1997) for a discussion of the complementary nature of the different approaches.

Table 8.1 Comparing the five approaches[1]

	Human Sociobiology	Human behavioural ecology	Evolutionary psychology	Memetics	Gene–culture coevolution (Dual-inheritance theory)
Level of explanation	Behaviour	Behaviour	Psychological mechanisms	Memes	Gene–meme combinations
Hypothesis generation	Gene's-eye view reasoning	Optimality models	Inference from evolutionary theory or history	Meme's-eye view reasoning	Mathematical models
Hypothesis testing methods	Multiple, but mainly ethnographic information	Quantitative ethnographic information	Multiple, but mainly questionnaires, lab experiments	Potentially multiple, including laboratory experiments	Mathematical modelling and simulation

Table 8.1 (*Contd*)

	Human Sociobiology	Human behavioural ecology	Evolutionary psychology	Memetics	Gene–culture coevolution (Dual-inheritance theory)
Comparator[2]	Multiple: Pleistocene hominids, primates, animal societies, optimality models	Optimality models	Pleistocene hominids	Genes, viruses	None
Is behaviour adaptive?	Yes	Yes	Not always, because of adaptive lag	Not always, because of parasitic memes	Usually, but cultural evolution renders maladaptive outcomes possible

Table 8.1 (*Contd*)

	Human Sociobiology	Human behavioural ecology	Evolutionary psychology	Memetics	Gene–culture coevolution (Dual-inheritance theory)
What is culture?	Multiple: cultural universals, behaviour elicited by ecological conditions, transmitted information	Multiple, but mainly behaviour elicited by ecological conditions	Multiple, but mainly cultural universals constrained by human nature	Socially transmitted information	Socially transmitted information guided by learning biases
What are human beings?	Sophisticated animals	Sophisticated animals characterized by extreme adaptability	Sophisticated animals guided by psychological adaptations	Sophisticated animals manipulated by cultural parasites	Sophisticated animals guided by genetic and cultural information

[1] Based on Smith, 2000.

[2] The entity with which comparisons are made to generate explanatory hypotheses.

Level of explanation

At first sight, there would appear to be characteristic differences between the schools in their level of explanation. Researchers frequently appear to be talking about different entities. Human sociobiologists tended to provide explanations at the level of behaviour. Human activities and strategies were described as adaptive traits, a perspective with which most human behavioural ecologists concur. For evolutionary psychologists, on the other hand, the principle focus of interest, and the level at which natural selection is deemed to operate is the psychological, with evolved cognitive mechanisms regulating behavioural outcomes. Meme enthusiasts state that the only psychological mechanism that really counts is the capacity for imitation (and other forms of social learning), and maintain that it is infectious cultural information that directs human behaviour. They argue that, with social learning, a new form of evolution at the cultural level took off, leaving biological evolution in its wake. Advocates of gene–culture coevolution also emphasize culturally transmitted information and agree with memeticists that transmitted culture plays a critical role in explanations for human behaviour. However, they also hold the view, common to many sociobiologists, behavioural ecologists, and evolutionary psychologists, that meme enthusiasts have underestimated the extent to which natural selection has fashioned human minds to structure how and what we learn.

However, a closer inspection reveals that these distinctions are more a question of focus than any fundamental differences of opinion. While human behavioural ecologists rarely describe their models as comprising psychological mechanisms, these analyses explicitly incorporate decision

rules and make assumptions as to the kinds of proximate
cues that are being attended to and the sources of information that are accrued. In a very real sense they are constructing models of psychological mechanisms, although
such models are framed in a different language to that of
evolutionary psychology. A number of articles by human
behavioural ecologists have explicitly dwelt on the possibility of adaptations at the psychological level (e.g. Draper,
1989; Turke, 1990; Borgerhoff Mulder, 1998; Smith, 2000).
Neither do evolutionary psychologists ignore behaviour. All
of the case studies that we describe in Chapter 5 have made
specific predictions as to the behavioural outcomes of
psychological mechanisms. Moreover, many evolutionary
psychologists envisage circumstances under which humans
may best be characterized as flexible adaptive strategists
(Buss, 1999). Furthermore, memetics and gene–culture
coevolutionary theory have been explicitly concerned with
the psychological mechanisms that underpin social learning. Indeed, perhaps the principal focus of the classic
gene–culture texts of Cavalli-Sforza and Feldman (1981)
and Boyd and Richerson (1985) is the question 'Through
which processes do humans learn from others?'. One recent
welcome development in memetics is the investigation of
meme–gene coevolution (Blackmore, 1999), which also
suggests a shift towards a consideration of other levels of
explanation. In short, while different practitioners place
more or less emphasis on psychological mechanisms,
behaviour, and cultural information, virtually all would
envisage some role for each, and there would seem to be
little of substance that necessarily separates the schools of
thought here.

Eric Alden Smith forcefully makes the same point in a
recent article:

Most evolutionary social scientists and biologists would
agree that complete evolutionary explanations of behavior
will include (i) heritable information that helps build
(ii) psychological mechanisms, which in turn produce
(iii) behavioural responses to (iv) environmental stimuli,
resulting in (v) fitness effects... (Smith, 2000, p. 35)

Smith notes that evolutionary psychology focuses on
(ii) psychological mechanisms and their links to (iii) behaviour and (iv) the environment; human behavioural ecology
focuses on (iii) behavioural responses with attention to
(iv) environmental stimuli and (v) fitness effects; while
gene–culture coevolution focuses on (i) genetic and cultural inheritance and its links to (ii) psychological mechanisms and (v) reproductive success.[3] Smith does not include
sociobiology or memetics in this analysis, but we could add
that classical sociobiology focused on (iii) behaviour and its
relationship to (i) genes and (v) fitness, while memetics is
concerned with (i) the cultural component of heritable
information, which is expressed in (iii) behaviour. Here
then, we are in agreement with Smith, who states that,
'Viewed in this light, a tentative case can be made for
explanatory complementarity' (2000, p. 35).

Hypothesis generation

Let us turn to the means by which the different approaches
generate hypotheses. Are there fundamental differences of
methodology between the schools?

Human sociobiologists generally employed gene's-eye
view reasoning to generate hypotheses about adaptive

---

[3] Note that Smith (2000) refers to gene–culture coevolution as dual-
inheritance theory (or DIT).

306 SENSE AND NONSENSE

human behaviour. While many such hypotheses were little more than speculation, others were tested by recourse to ethnographic data, through experimentation, or in other ways. Some sociobiologists suggested that human behaviour was best interpreted through comparison with particular social groups, such as hunter–gatherers, primates, or social animals, while others regarded humans as flexibly adapted to local ecological conditions. The latter school grew into human behavioural ecology, whose practitioners use optimality models to produce predictions concerning human behaviour on the assumption that it is adaptive. These predictions are generally tested with quantitative ethnographic data from non-Western, preindustrial societies, such as the Ache of Paraguay or the Canadian Inuit.

Evolutionary psychologists employ inferences from evolutionary theory and knowledge of our past as Pleistocene hunter–gatherers to generate hypotheses about the evolved psychological mechanisms that underpin human behaviour. Such hypotheses are generally tested with questionnaires, with laboratory experiments, or by analysing existing data records (such as crime statistics).

Memeticists draw on analogies with genes and viruses and employ meme's-eye view reasoning to generate hypotheses about how people's thoughts, behaviour, and institutions are manipulated by infectious, socially transmitted information. While memetics provides a stimulating new way to think about human behaviour, it has yet to develop a rigorous empirical programme, although there are signs that this is changing and a number of methods have been proposed (see Chapter 6).

Gene–culture coevolution is admired by many and practised by few, largely because it too has spawned little empirical research. However, once again there is no reason why this has to be the case, as the empirical methods described in Chapters 6 and 7 outline. In contrast to memetics, however, gene–culture coevolution has developed a rigorous theoretical tradition. Advocates construct mathematical models that describe how two streams of inherited information, genes and culture, interact to generate human behaviour.

In Chapter 4, we suggested that much of the dispute between human behavioural ecologists and evolutionary psychologists could be put down to their interest in different aspects of the evolutionary process. In simple terms, human behavioural ecologists are more concerned with whether human behaviour is *adaptive* (that is, whether it varies in response to current conditions so as to enhance reproductive success), while evolutionary psychologists primarily endeavour to isolate human *adaptations* (that is, characters favoured by natural selection for their effectiveness in a particular role). Not only are adaptations and adaptive behaviour not the same, but they can be regarded as orthogonal or independent. We illustrated this in Figure 4.1, where we distinguished between four products of the evolutionary process (current adaptations, past adaptations, exaptations, and dysfunctional by-products). When depicted in this manner, the complementary nature of the two leading evolutionary approaches is again highlighted. Neither assessing whether a human character is adaptive nor isolating human adaptations is sufficient alone to provide a satisfactory evolutionary explanation for the characteristic in focus. Both approaches are necessary if the characteristic is to be fully understood as a product of evolution, with its function and history each well understood.

More generally, it would seem that researchers interested in human behaviour and evolution have been interested in four different types of question concerning a particular character, each of them entirely legitimate. Human behavioural ecologists (and some human sociobiologists) posit the question 'Is the character adaptive?' In contrast, evolutionary psychologists (and other sociobiologists) ask 'Is the character an adaptation?' Meme enthusiasts are interested in a parallel question, namely 'Is the character an adaptation-like product of cultural evolution?' Finally, gene–culture coevolutionary researchers are interested in a fourth type of question, namely 'How do characters evolve?' The empirical methods for memetics and gene–culture coevolution described in the preceding chapters also focus on the characteristics of the process of evolution. It would seem then that researchers are simply asking different kinds of evolutionary questions, but that all are important and will generate complementary knowledge.

Hypothesis testing methods

Does the manner in which researchers in different schools test their hypotheses reflect any essential differences of opinion as to the best means to do science? We think not. There is little doubt that many of the suture lines along which the field is dissected reflect historical accidents. The most obvious reason why human behavioural ecologists test their hypotheses with ethnographic data from naturalistic environments, while evolutionary psychologists distribute questionnaires to undergraduate students, and gene–culture enthusiasts derive systems of recursive equations, is that anthropology, psychology, and population genetics have long traditions of testing their hypotheses in such ways.

Yet these alternative means of assessment are to a large extent arbitrary. There is nothing to stop evolutionary psychologists from constructing mathematical models[4] or to prevent human behavioural ecologists from exploring the proximate constraints on adaptive behaviour with psychological experiments. According to Daly and Wilson (2000), there is already considerable overlap in methodology between evolutionary psychologists and human behavioural ecologists. Equally, there is no reason why memeticists couldn't carry out field experiments on meme diffusion or why advocates of gene–culture coevolution couldn't make quantitative predictions about adaptive and (perhaps more interestingly) maladaptive behaviour. Indeed, some of the most exciting work to be done is precisely that at the boundaries between approaches. Genuine advances will be made when, for instance, human behavioural ecologists build models with a transmitted culture or when memeticists start wondering how evolved psychological mechanisms affect an individual's susceptibility to acquiring memes.

At the methodological level then, the approaches exploit alternative forms of data collection and hypothesis testing, and there are considerable differences of focus and emphasis. However, there is nothing incompatible about these methodological differences and nothing to stop researchers from using all of the available tools.

What is culture?

Historically, the notion of culture has proven very difficult for academics to pin down. In a famous article published in 1952, two prominent anthropologists identified 164 different definitions of culture proposed by social scientists

---

[4] See for instance, Miller and Todd's (1995) analysis of mate choice and sexual selection.

(Kroeber and Kluckholm, 1952), and that number has undoubtedly grown. While there is far from a consensus even today, most social scientists would agree on two points, that culture is composed of symbolically encoded acquired information and that it is socially transmitted within and between populations, largely free of biological constraints. Is that the way evolutionists regard culture? For the most part it would seem not.

As we have stressed, human sociobiology was a broad church, encompassing researchers with an array of different views, including views of culture. In most cases, human culture was regarded as little different from any other aspect of the human phenotype. Yet for some sociobiologists, culture was behaviour elicited by ecological conditions (Alexander, 1979), while for others (e.g. Wilson, 1975) it was composed largely of cultural universals closely tied to our biological nature. An additional perspective that was entertained by Wilson's human sociobiology was that genetic diversity underlies the differences between cultures (Wilson, 1975; Lumsden and Wilson, 1981). This was one of the most controversial aspects of human sociobiology and a primary source of hostility from the social science community (Segerstråle, 2000). However, this view does not find favour among any of the contemporary evolutionary schools. All modern approaches accept that cultural change can occur without accompanying genetic change and that individuals from genetically distinct populations acquire the cultural traits of each other without difficulty (Flinn, 1997). It is worth reiterating this point because it remains a source of misunderstanding among critics of evolutionary perspectives on culture.

The latter apart, the principal sociobiological perspectives are, to some extent, represented by the contemporary evolutionary approaches of human behavioural ecology and evolutionary psychology. The human behavioural ecology tradition tends to regard the variability in human society and culture as entities that are, to a large extent, evoked by the ecological environment. That is not to say that culture is not learned nor socially transmitted but rather that humans are predisposed to learn that which maximizes their inclusive fitness by satisfying various proximal goals, such as to obtain food and mates, and to avoid danger and disease. Culture is part of the unusually broad and flexible evolved mechanisms of behavioural adaptation that characterize humans, such that we are adapted to a wide range of conditions rather than to a particular environmental state.

In contrast, evolutionary psychologists are most interested in those aspects of human culture that are found universally among all peoples, our human nature. According to this view, human minds are organized by complex, evolved information processing structures that channel learning aptitudes towards that which was beneficial in our environment of evolutionary adaptedness. Compared to human behavioural ecologists, evolutionary psychologists have a view of the mind that is much more structured and prespecified by our genetic heritage and perhaps less flexible in the face of environmental variability. As a consequence, evolutionary psychologists anticipate that a larger proportion of human behaviour will be maladaptive than human behavioural ecologists. It would only be fair to point out, however, that most evolutionary psychologists recognize some evoked culture and most human behavioural ecologists accept that there are cultural universals.

A third evolutionary take on culture, represented by the school of memetics, is that it is a dynamic evolutionary system in its own right.[5] Culture is phenotypic plasticity that acquired its own intrinsic capacity to change and is now out of genetic control. We don't expect a flu virus to operate to our advantage, so why should we expect a 'mind virus' always to be in our interests? For meme advocates, not only is cultural evolution largely unconstrained by genetic predispositions, but genetic evolution may itself be driven by cultural imperatives (Blackmore, 1999).

A fourth evolutionary style asserts that cultural information is socially transmitted between individuals but that its acquisition is biased by evolved learning rules and motivational priorities. This is the perspective of much modern gene–culture coevolution. Advocates of memetics and gene–culture coevolution differ from most evolutionary psychologists and human behavioural ecologists in believing that cultural phenomena cannot be fully understood without recourse to the intrinsic processes of cultural change, which are at least partly independent of the processes of biological evolution. Such cultural processes do not necessarily act to maximize fitness and hence may result in arbitrary or even maladaptive outcomes. These researchers agree that cultural processes have a historical dependency, such that culture cannot be well predicted without knowledge of the population's traditions. In other words, there are processes besides evolutionary design of cognitive architecture that affect culture content. However, gene–culture enthusiasts typically attribute more importance to evolved learning biases than memeticists. For instance, Boyd and Richerson's model of 'direct bias' can

---

[5] This perspective can also be found among some practitioners of cultural evolution theory (e.g. Cavalli-Sforza and Feldman, 1981).

represent situations in which individuals evaluate the relative success of two or more alternative traits and choose the option that suits their genotype.

Evolutionary psychology would seem to have greater conflict with memetics, which denies any substantive filtering role for evolved psychological mechanisms, than with gene–culture coevolution, which sees evolved predispositions frequently being instrumental in decision-making. However, evolutionary psychology's focus on genetically evolved panhuman cognitive algorithms tends to make its practitioners relatively hostile to the view that humans acquire a substantial proportion of their beliefs and preferences through cultural inheritance and that these can change through cultural evolution (Smith, 2000).

It may seem surprising that such a diversity of views should be found amongst researchers, all of whom share the belief that human learning mechanisms have been moulded by natural selection. Surely if our capacity for culture is an adaptation, we should expect social learning to be adaptive too? However, this problem is more complicated than it first appears. Using our evolved *capacity* for social learning, we could theoretically acquire maladaptive *information*, perhaps because it is outdated, or only adaptive in other environments or for other people (Boyd and Richerson, 1985). Yet it does not follow that maladaptive cultural information will inevitably be expressed in maladaptive *behaviour*, as individuals may filter out what is inappropriate to them or adjust their behaviour in the light of negative experiences (Galef, 1995). It is even possible for maladaptive transmitted information to be expressed in adaptive behaviour in cases where it pays to do what others are doing, even if the cultural *traditions* are suboptimal (Laland and Williams, 1998). The capacity for social learning, the socially transmitted information, the expressed learned behaviour, and the population's behavioural tradition are four separate entities, and the demonstration that one is adaptive or maladaptive does not constitute evidence that others are likewise. In the light of this complexity, it is easier to envisage how disparate views are possible.

Here then is a genuine ideological difference between the schools. At the core of these distinctions are different views as to how humans learn from each other. It turns out that social learning is the key process underlying these evolutionary paradigms (Flinn, 1997). Are we predisposed to learn what is currently adaptive, guided by proximate motivational cues such as hunger or fear as the human behavioural ecologists maintain? Or is our brain set up to prioritize learning that which was important in the past, as the evolutionary psychologists suspect? Do we acquire whatever behaviour or information just happens to be easiest or most compelling to learn, as memeticists would have it? Or is our learning dependent partly on evolved predispositions and partly on cultural processes, as the gene–culture coevolution theorists have it? In fact, it is not inconceivable that all these perspectives could be correct to some degree. That is, each of these views could be true for different learned behaviour patterns or on different occasions (Smith, 2000). The question then turns to how frequently each finds empirical support.

Ultimately, these issues will have to be worked out through experimentation and other forms of research rather than through the polemical pronouncements that have thus far dominated most attempts to confront them (Smith, 2000). What kind of studies would tell us which of these views of culture is closest to the mark? There are two obvious places to start. First, researchers could carry out quantitative analyses across a multitude of behavioural traits to measure to what extent, or on what percentage of occasions, human behaviour is currently adaptive. A good example of this type of study was carried out by anthropologist Robert Aunger on the food preferences of horticulturists and Pygmy foragers living in the Ituri forest of the Democratic Republic of Congo (Aunger, 1992; 1994a, b). Aunger observed that different populations varied as to which foodstuffs they exploited and which they avoided and asked whether their food avoidances were maladaptive. He found that individuals in one of the four ethnic groups suffered a selective disadvantage that resulted from their cultural beliefs about food. These maladaptive food avoidances generally reduced fitness by a few per cent, mainly through compromising female fertility. Aunger interprets these cases as likely to have been the outcome of cultural processes, rather than a consequence of adaptive lag. If Aunger's data can be regarded as representative, then a significant minority of cases of human behaviour are maladaptive. It is, of course, conceivable that the amount of maladaptive behaviour will differ in other societies or in relation to other domains.

A second, equally informative kind of analysis would measure, among diverse traits and across a broad range of populations, what percentage of the variance in behaviour is explained by local ecology and what percentage is better predicted by cultural history. In the previous chapter, we described just such an analysis, carried out by Guglielmino et al. (1995) on 277 African societies, which found that most traits correlated with cultural history rather than with ecology. If this study is representative, then socially trans-mitted cultural traditions are a lot more important than most evolution-minded researchers envisage. However, this finding can still be reconciled with the view that most human behaviour is adaptive. First, it may be that cultural history specifies which, among several broadly adaptive alternative behaviour patterns, is preferred by a given population at a given time. Secondly, experimental studies of social learning in animals have shown that it may be adaptive for an individual to conform with what the majority are doing, even where the tradition is suboptimal, if there are fitness costs associated with going it alone, for instance if an isolated forager is more vulnerable to predation (Laland and Williams, 1998). Conformity is likely to be even more important among human populations than in animals (Boyd and Richerson, 1985; Henrich *et al.*, 2001).

We began this section with the observation that most social scientists regard culture as symbolically encoded acquired information that is socially transmitted between individuals unbound from biological constraints. Studies such as Guglielmino *et al.* suggest that evolutionary minded researchers may have underestimated the amount and significance of transmitted culture. A rejection of cultural determinism and of a *tabula rasa* model of the mind does not have to encompass a neglect of cultural processes. Ultimately, the findings of biology and social science will need to be compatible if either are to rate as satisfactory.

Conclusions

Given the diverse backgrounds and interests of the practitioners, it is hardly surprising that several distinct perspectives on human behaviour have emerged, largely reflecting the methodological and conceptual habits of the parent disciplines. Inevitably, the different approaches are sometimes seen as providing competing views of human behaviour. However, when these alternatives are examined more closely it becomes clear that there is little that is genuinely incompatible about their explanations or methodologies. While there are some theoretical differences, these will eventually be settled by empirical research. In the mean time, as our case study on infanticide demonstrates, there is a complementarity of information generated by the different methods. As Hinde's analysis of war illustrates, individual researchers are free to draw from different styles to synthesize their own integrative and pluralistic evolutionary analyses.

Neither should we be surprised if differences of opinion emerge as to the best way forward in a young and provisional science. As we have seen, squabbles have broken out between the different schools and occasionally the exchanges have become quite heated. The last twenty-five years have witnessed repeated pleas in the human behaviour and evolution literature for the 'rival factions' to settle their differences, to dwell on their common ground, and to bond in the face of the considerable hostility that is seen as emanating from the massed ranks of the social sciences. Yet the mere assertion that 'we're all in this together' is unlikely to generate a true integration of perspective when particular camps regard the alternatives as out of touch with relevant literatures or methodologically weaker than themselves. We do not agree with the view that it is divisive to dwell on any differences of methodology or conviction within the field and even that it is damaging for the discipline to be exacting. On the contrary, to the extent that human sociobiology and its descendants have consciously or inadvertently discouraged a self-critical ethos among their practitioners, we regard this as a barrier to integration. Why should researchers want the respectable, disciplined work of their subdiscipline to be associated with the sensationalist claims and superficial analyses of other researchers? Inevitably they will fear that inflammatory declarations, careless popularizations, and adaptationist story-telling will produce a backlash against all evolutionary approaches in the social sciences.[6] Given the valuable research on human behaviour and evolution carried out by all the schools, this would not only be a great shame but an unnecessary tragedy. The field of evolution and human behaviour is no longer a vulnerable sapling, but has developed into a vibrant and vigorously growing tree with roots sufficiently well established for it to be able to stand up to, and indeed benefit from, healthy pruning.

A delicate balance must be struck here. While there is a need for genuine pluralism of methodology, it does not follow that all analyses based loosely on evolutionary conjecture are salutary. High standards of research, rather than unconditional positive regard to fellow enthusiasts, are the best defence against external disapprobation. What is needed is a pluralistic yet rigorous, fertile yet self-critical, scientific discipline that at the same time champions *bona fide* evolutionary methods and inferences but clamps down hard on undisciplined story-telling and potentially damaging or abusive evolutionary reasoning. A genuine marriage of the biological and social sciences will only emerge when the ratio of sense to nonsense is improved.

---

6   Precisely these fears are expressed by Smith *et al.* (2001).

Further reading

Chapter 1 – Sense and nonsense

For recent discussions of the nature–nurture debate, see Bateson and Martin's *Design for a Life: How Behaviour Develops* (1999) and Lewontin's *The Triple Helix* (2000). Introductions to modern evolutionary theory are provided by Ridley's *Evolution* (1997) and Futuyma's *Evolutionary Biology* (1998). Jared Diamond's *The Rise and Fall of the Third Chimpanzee* (1991) and Paul Ehrlich's *Human Natures. Genes, Cultures and the Human Prospect* (2000) provide readable overviews that touch on evolutionary theories of human behaviour.

Chapter 2 – A history of evolution and human behaviour

For Darwin on human behaviour see *The Descent of Man* (1871; reprinted 1981) and *The Expression of the Emotions in Man and Animals* (1872; reprinted 1998). For a modernized version of *The Origin of Species*, read Steve Jones's *Almost Like a Whale: the Origin of Species Updated* (1999). For more information on the history of evolution and the study of behaviour, see Boakes's *From Darwin to Behaviourism* (1984), Oldroyd's *Darwinian Impacts* (1983), and Plotkin's *Evolution in Mind* (1997). For an anthropologist's view on these issues, see Kuper's *The Chosen Primate* (1994). Johan Bolhuis and Jerry Hogan's *The Development of Animal Behaviour* (1999) collates some landmark essays on behavioural development by Lehrman, Kuo, Lorenz, Hinde, and others.

Chapter 3 – Human sociobiology

For introductions to gene's-eye view thinking, read Williams's *Adaptation and Natural Selection: a Critique of Some Current Evolutionary Thought* (1966) and Dawkins's *The Selfish Gene* (1976). To find out why Wilson aroused such controversy, read his final chapter in *Sociobiology* (1975) and *On Human Nature* (1978). For an excellent sociological analysis of the sociobiology debate, see Segerstråle's *Defenders of the Truth: the Sociobiology Debate* (2000). For a personal account of the sociobiology debate see Wilson's autobiography *Naturalist* (1994). For hostile views of human sociobiology see Sahlins's *The Use and Abuse of Biology: an Anthropological Critique of Sociobiology* (1976) and Kitcher's *Vaulting Ambition* (1985)

Chapter 4 – Human behavioural ecology

For recent collections of empirical and theoretical works, see *Adaptation and Human Behavior: an Anthropological Perspective* (2000) edited by Cronk, Chagnon, and Irons, and *Evolutionary Ecology and Human Behavior* (1992), edited by Smith and Winterhalder. *Human Nature: a Critical Reader* (1997), edited by Laura Betzig, reproduces many of the key papers in the human behavioural ecology and evolutionary psychology literature. For an example of detailed data collection and life history analyses, see *Ache Life History: the Ecology and Demography of a Foraging People* (1996) by Hill and Hurtado.

Chapter 5 – Evolutionary psychology

Steven Pinker's *How the Mind Works* (1997) and Henry Plotkin's *Evolution in Mind* (1997) constitute readable overviews of evolutionary psychology. Barret *et al.*'s *Human Evolutionary Psychology* (2001) is a detailed student textbook, but with a commendably broad perspective. Barkow *et al.* (1992) contains a series of evolutionary psychology articles in the Santa Barbara tradition. Donald Symons's (1987, 1989, 1990) essays draw out the distinction between evolutionary psychology and human behavioural ecology.

Chapter 6 – Memetics

Susan Blackmore's *The Meme Machine* (1999) provides an accessible introduction to memetics. Dennett's *Darwin's Dangerous Idea: Evolution and the Meaning of Life* (1995) and *Consciousness Explained* (1991) provide a philosopher's views on memes. A variety of views on memetics are contained in *Darwinizing Culture: the Status of Memetics as a Science* (Aunger, 2000). For ideas on potential empirical methods for detecting selection, read Endler's *Natural Selection in the Wild* (1986). For a review of the bird song literature, see Catchpole and Slater's book *Bird Song* (1995).

Chapter 7 – Gene–culture coevolution

A comprehensive introduction to the methods and findings of gene–culture coevolution can be found in Boyd and Richerson's *Culture and the Evolutionary Process* (1985), and Cavalli-Sforza and Feldman's *Cultural Transmission and Evolution: A Quantitative Approach* (1981). For a mathematical account of some key methods see Feldman and Cavalli-Sforza's (1976) paper. A more accessible overview is provided in the paper by Feldman and Laland (1996), while Laland *et al.* (1995a) provides a worked example of the mathematical theory. Paul Ehrlich's *Human Natures. Genes, Cultures and the Human Prospect* (2000) provides a readable overview that is sympathetic to gene–culture co-evolution.

Chapter 8 – Comparing and integrating approaches

For a textbook which provides a broad perspective on evolution and human behaviour, see Barrett *et al.* (2001). The integration of approaches is discussed by Smith (2000). Flinn (1997) also compares different evolutionary approaches and provides an alternative perspective on this debate. For a discussion of war, see the collected papers in Hinde and Watson (1995).

References

Adenzato, M. 2000. Gene–culture coevolution does not replace standard evolutionary theory. *Behavioral and Brain Sciences* 23: 146–7.

Alexander, R. D. 1974. The evolution of social behavior. *Annual Review of Ecology and Systematics* 5: 325–83.

Alexander, R. D. 1979. *Darwinism and Human Affairs.* London: Pitman.

Alexander, R. D. and Noonan, K. M. 1979. Concealment of ovulation, parental care, and human social evolution. In: *Evolutionary Biology and Human Social Behavior: an Anthropological Perspective.* Ed. N. A. Chagnon and W. Irons. Duxbury Press. Pp. 436–53.

Allen, E. *et al.* 1975. Letter. *New York Review of Books.* 13 Nov, 182, 184–6.

Aoki, K. 1986. A stochastic model of gene–culture coevolution suggested by the 'culture historical hypothesis' for the evolution of adult lactose absorption in humans. *Proceedings of the National Academy of Sciences USA* 83: 2929–33.

Aoki, K. and Feldman, M. W. 1987. Toward a theory for the evolution of cultural communication: coevolution of signal transmission and reception. *Proceedings of the National Academy of Sciences USA* 84: 7164–8.

Aoki, K. and Feldman, M. W. 1989. Pleiotropy and pre-adaptation in the evolution of human language capacity. *Theoretical Population Biology* 35: 181–94.

Aoki, K. and Feldman, M. W. 1991. Recessive hereditary deafness, assortative mating, and persistence of a sign language. *Theoretical Population Biology* 39: 358–72.

Aoki, K. and Feldman, M. W. 1997. A gene–culture coevolutionary model for brother–sister mating. *Proceedings of the National Academy of Sciences USA* 94: 13046–50.

Aoki, K., Shida, M. and Shigesada, N. 1996. Travelling wave solutions for the spread of farmers into a region occupied by hunter–gatherers. *Theoretical Population Biology* 50: 1–17.

Ardrey, R. 1966. *The Territorial Imperative*. London: Collins.

Aunger, R. 1992. The nutritional consequences of rejecting food in the Ituri forest of Zaire. *Human Ecology* 30: 1–29.

Aunger, R. 1994a. Are food avoidances maladaptive in the Ituri Forest of Zaire? *Journal of Anthropological Research* 50: 277–310.

Aunger, R. 1994b. Sources of variation in ethnographic interview data: food avoidances in the Ituri Forest, Zaire. *Ethnology* 33: 65–99.

Aunger, R. 2000. *Darwinizing Culture: the Status of Memetics as a Science*. Oxford: Oxford University Press.

Aunger, R. 2000a. Introduction. In: *Darwinizing Culture: the Status of Memetics as a Science*. Oxford: Oxford University Press. Pp. 1–23.

Aunger, R. 2000b. Conclusion. In: *Darwinizing Culture: the Status of Memetics as a Science*. Oxford: Oxford University Press. Pp. 205–32.

Aunger, R. 2000c. The life history of culture learning in a face-to-face society. *Ethos* 28: 1–38.

Aunger, R. 2002. *The Electric Meme: a New Theory of How We Think and Communicate.* New York: The Free Press.

Bagemihl, B. 1999. *Biological Exuberance: Animal Homosexuality and Natural Diversity.* London: Profile Books.

Barkow, J. H., Cosmides, L., and Tooby, J. 1992. *The Adapted Mind: Evolutionary Psychology and the Generation of Culture.* Oxford: Oxford University Press.

Barrett, L., Dunbar, R., and Lycett, J. 2001. *Human Evolutionary Psychology.* London: Macmillan.

Bateson P. P. G. 1981. Sociobiology and genetic determinism. *Theoria to Theory* 14: 291–300.

Bateson P. P. G. 1994. The dynamics of parent–offspring relationships in mammals. *Trends in Ecology and Evolution* 9: 399–402.

Bateson, P. 1996. Design for a life. In: *The Lifespan Development of Individuals.* Ed. D. Magnusson. Cambridge: Cambridge University Press. Pp. 1–20.

Bateson, P. and Martin, P. 1999. *Design for a Life: How Behaviour Develops.* London: Jonathan Cape.

Betzig, L. (ed.). 1997. *Human Nature: a Critical Reader.* Oxford: Oxford University Press.

Binmore K. 1998. *Game Theory and the Social Contract: Just Playing. Vol. 2.* Cambridge, MA: MIT Press.

Blackmore, S. 1999. *The Meme Machine.* Oxford: Oxford University Press.

Blackmore, V. and Page, A. 1989. *Evolution: the Great Debate.* Oxford: Lion Publishing.

Blackwell, A. B. 1875. *The Sexes Throughout Nature.* New York: G. P. Putnam.

Bloch, M. 2000. A well-disposed social anthropologist's problems with memes. In: *Darwinizing Culture: the Status of Memetics as a Science.* Oxford: Oxford University Press. Pp. 189–203.

Blurton Jones, N. 1986. Bushman birth spacing: a test for optimal interbirth interval. *Ethology and Sociobiology* 7: 91–105.

Blurton Jones, N. 1994. A reply to Dr. Harpending. *American Journal of Physical Anthropology* 93: 391–7.

Blurton Jones, N. 1997. Too good to be true? Is there really a trade-off between number and care of offspring in human reproduction? In: *Human Nature: a Critical Reader.* Ed. L. Betzig. Oxford: Oxford University Press. Pp. 83–6.

Blurton Jones, N. and Sibly, R. M. 1978. Testing adaptiveness of culturally determined behaviour: do bushman women maximise their reproductive success by spacing births widely and foraging seldom? In: *Human Behaviour and Adaptation.* Ed. N. Blurton Jones and V. Reynolds. London: Francis Taylor. Pp. 135–58.

Boakes, R. 1984. *From Darwin to Behaviourism: Psychology and the Minds of Animals.* Cambridge: Cambridge University Press.

Bolhuis, J. J. and Hogan, J. A. 1999. *The development of Animal Behavior. A Reader.* Oxford: Blackwell.

Bolhuis, J. J. and MacPhail, E. M. 2001. A critique of the neuroecology of learning and memory. *Trends in Cognitive Sciences* 5: 426–433.

Bonner, J. T. and May, R. M. 1981. Introduction. In: Darwin C. *The Descent of Man, and Selection in Relation to Sex.* Princeton: Princeton University Press.

Borgerhoff Mulder, M. 1990. Kipsigis women's preference for wealthy men: evidence for female choice in mammals? *Behavioral Ecology and Sociobiology* 27: 255–64.

Borgerhoff Mulder, M. 1991. Human behavioural ecology. In: *Behavioural Ecology: an Evolutionary Approach.* Ed. J. R. Krebs and N. B. Davies. Oxford: Blackwell Scientific Publications.

Borgerhoff Mulder, M. 1998. The demographic transition: are we any closer to an evolutionary explanation? *Trends in Ecology and Evolution* 13: 266–70.

Borgerhoff Mulder, M. 2000. Optimizing offspring: the quantity–quality tradeoff in agropastoral Kipsigis. *Evolution and Human Behavior* 21: 391–410.

Borgerhoff Mulder, M., Richerson, P. J., Thornhill, N. W., and Voland, E. 1997. The place of behavioral ecological anthropology in evolutionary social science. In: *Human by Nature: Between Biology and the Social Sciences.* Ed. P. Weingart, S. D. Mitchell, P. J. Richerson, and S. Maasen. New Jersey: Erlbaum. Pp. 253–82.

Bouchard, T. J. and McGue, M. 1981. Familial studies of intelligence. *Science* 212: 1055–9.

Bouchard, T. J. *et al.* 1990. Sources of human psychological differences: the Minnesota study of twins reared apart. *Science* 250: 223–8.

Bowlby, J. 1969. *Attachment and Loss: Volume 1 Attachment.* London: Hogarth Press.

Bowles, S. 2000. Economic institutions as ecological niches. *Behavioral and Brain Sciences* 23: 148–9.

Boyd, R. and Richerson, P. J. 1982. Cultural transmission and the evolution of cooperative behavior. *Human Ecology* 10: 325–51.

Boyd, R. and Richerson, P. 1983. The cultural transmission of acquired variation: effects on genetic fitness. *Journal of Theoretical Biology* 100: 567–96.

Boyd, R. and Richerson, P. J. 1985. *Culture and the Evolutionary Process.* Chicago: Chicago University Press.

Boyd, R. and Silk, J. B. 1997. *How Humans Evolved.* New York: Norton.

Brase, G. L., Cosmides, L., and Tooby, J. 1998. Individuation, counting, and statistical inference: the role of frequency and whole-object representations in judgment under uncertainty. *Journal of Experimental Psychology: General* 127: 3–21.

Brodie, R. 1996. *Virus of the Mind: The New Science of the Meme.* Seattle, WA: Integral Press.

Brown, D. E. 1991. *Human Universals.* New York: McGraw-Hill.

Brown, G. R. 2000. Can studying non-human primates inform us about human rape? A zoologist's perspective. *Psychology, Evolution and Gender* 2: 321–324.

Brown, G. R. 2001. Sex-biased investment in nonhuman primates: can Trivers and Willard's theory be tested? *Animal Behaviour* 61: 683–94.

Bull, L., Holland, O., and Blackmore, S. 2001. On meme–gene coevolution. *Artificial Life* 6: 227–35.

Burkhardt, R. W. 1983. The development of an evolutionary ethology. In: *Evolution from Molecules to Men.* Ed. D. S. Bendall. Cambridge: Cambridge University Press.

Burley, N. 1979. The evolution of concealed ovulation. *American Naturalist* 114: 835–58.

Burnet, F. M. 1959. *The Clonal Selection Theory of Acquired Immunity*. Nashville: Vanderbilt University Press.

Burt, A. 1992. 'Concealed ovulation' and sexual signals in primates. *Folia Primatologica* 58: 1–6.

Buss, D. M. 1994. *The Evolution of Desire: Strategies of Human Mating*. New York: HarperCollins.

Buss, D. M. 1995. Evolutionary psychology: a new paradigm for psychological science. *Psychological Inquiry* 6: 1–30.

Buss, D. M. 1999. *Evolutionary Psychology: the New Science of the Mind*. London: Allyn and Bacon.

Buss, D. M. *et al*. 1990. International preferences in selecting mates: a study of 37 cultures. *Journal of Cross-Cultural Psychology* 21: 5–47.

Campbell, D. T. 1974. Evolutionary epistemology. In: *The Philosophy of Karl R. Popper*. Ed. P. A. Schipp. La Salle, IL: Open Court.

Caro, T. M. and Borgerhoff Mulder, M. 1987. The problem of adaptation in the study of human behavior. *Ethology and Sociobiology* 8: 61–72.

Caro, T. M., Sellen, D. W., Parish, A., Frank, R., Brown, D. M., Voland, E., and Borgerhoff Mulder, M. 1995. Termination of reproduction in nonhuman and human female primates. *International Journal of Primatology* 16: 205–20.

Cartwright, J. 2000. *Evolution and Human Behaviour*. London: Macmillan.

Catchpole, C. K. and Slater, P. J. B. 1995. *Bird Song: Biological Themes and Variations*. Cambridge: Cambridge University Press.

Cavalli-Sforza, L. L. and Feldman, M. W. 1973. Models for cultural inheritance. I. Group mean and within group variation. *Theoretical Population Biology* 4: 42–55.

Cavalli-Sforza, L. L. and Feldman, M. W. 1981 *Cultural Transmission and Evolution: A Quantitative Approach.* Princeton : Princeton University Press.

Cavalli-Sforza, L. L., Feldman, M. W., Chen, K. H., and Dornbusch, S. M. 1982. Theory and observation in cultural transmission. *Science* 218: 19–27.

Chagnon, N. A. and Irons, W. 1979. *Evolutionary Biology and Human Social Behavior: an Anthropological Perspective.* North Scituate, MA: Duxbury Press.

Christenfeld, N. and Hill, E. 1995. Whose baby are you? *Nature* 378: 669.

Clarke, G. M. 1995. Relationships between developmental stability and fitness: applications for conservation biology. *Conservation Biology* 9: 18–24.

Clarke, J. I. 2000. *The Human Dichotomy: the Changing Numbers of Males and Females.* Oxford: Elsevier Science.

Clutton-Brock, T. H. and Parker, G. A. 1995. Sexual coercion in animal societies. *Animal Behaviour* 49: 1345–65.

Cosmides, L. 1989. The logic of social exchange: has natural selection shaped how humans reason? Studies with the Wason selection task. *Cognition* 31: 187–276.

Cosmides, L. and Tooby, J. 1987. From evolution to behavior: evolutionary psychology as the missing link. In: *The Latest on the Best: Essays on Evolution and Optimality.* Ed. J. Dupré. Cambridge MA: MIT Press.

Cosmides, L. and Tooby, J. 1992. Cognitive adaptations for social exchange. In: *The Adapted Mind: Evolutionary*

Psychology and the Generation of Culture. Ed. J. H. Barkow, L. Cosmides, and J. Tooby. Oxford: Oxford University Press. Pp. 163–228.

Coyne, J. A. and Berry, A. 2000. Rape as an adaptation: is this contentious hypothesis advocacy, not science? *Nature* 404: 121–122.

Cranach, M. von, Foppa, K., Lepenies, W., and Ploog, D. (eds.) 1979. *Human Ethology: Claims and Limits of a New Discipline.* Cambridge: Cambridge University Press.

Cronk, L. 1991. Human behavioral ecology. *Annual Review of Anthropology* 20: 25–53.

Cronk, L. 1995. Commentary. *Current Anthropology* 36: 147.

Cronk, L., Chagnon, N., and Irons, W. (eds) 2000. *Adaptation and Human Behavior: an Anthropological Perspective.* New York: Aldine de Gruyter.

Crook, J. 1964. The evolution of social organization and visual communication in the weaver birds (*Ploceinae*). *Behaviour* 10: 1–178.

Crook, J. 1965. The adaptive significance of avian social organization. *Symposia of the Zoological Society of London* 4: 181–218.

Crook, J. and Crook, S. J. 1988. Tibetan polyandry: problems of adaptation and fitness. In: *Human Reproductive Behaviour: a Darwinian Perspective.* Ed. L. Betzig, M. Borgerhoff Mulder and P. Turke. Cambridge: Cambridge University Press. Pp. 97–114.

Crook, J. and Gartlan, J. S. 1966. Evolution of primate societies. *Nature* 210: 1200–3.

Crosson, B., Moberg, P. J., Boone, J. R., Gonzalez Rothi, L. J., and Raymer, A. 1997. Category-specific naming deficit for medical terms after dominant thalamic/capsular hemorrhage. *Brain and Language* 60: 407–42.

Crow, J. F. 2001. The beanbag lives on. *Nature* 409: 771.

Daly, M. 1982. Some caveats about cultural transmission models. *Human Ecology* 10: 401–8.

Daly, M. and Wilson, M. I. 1982. Whom are newborn babies said to resemble? *Ethology and Sociobiology* 3: 69–78.

Daly, M. and Wilson, M. 1983. *Sex, Evolution and Behavior. 2nd edn.* Belmont CA: Wadsworth.

Daly, M. and Wilson, M. 1988. *Homicide.* New York: Aldine.

Daly, M. and Wilson, M. I. 1999. Human evolutionary psychology and animal behaviour. *Animal Behaviour* 57: 509–19.

Daly, M. and Wilson, M. 2000. Reply to Smith *et al. Animal Behaviour* 60: F27-F29.

Daniels, D. 1983. The evolution of concealed ovulation and self-deception. *Ethology and Sociobiology* 4: 69–87.

Darwin, C. 1859. *The Origin of Species by Means of Natural Selection, or the Preservation of Favoured Races in the Struggle for Life.* London: John Murray (1st edn repr. Penguin Books, London; 1968).

Darwin, C. 1871. *The Descent of Man and Selection in Relation to Sex.* London: John Murray (1st edn repr. Princeton University Press, Princton NJ; 1981).

Darwin, C. 1872. *The Expression of the Emotions in Man and Animals.* London: John Murray (3rd edn repr. Harper-Collins, London; 1998).

Davies, N. B. 1989. Sexual conflict and the polygamy threshold. *Animal Behaviour* 38: 226–34.

Dawkins, R. 1976. *The Selfish Gene.* Oxford: Oxford University Press.

Dawkins, R. 1981. Selfish genes in race or politics. *Nature* 289: 528.

Dawkins, R. 1982. *The Extended Phenotype*. Oxford: Oxford University Press.

Delius, J. D. 1991. The nature of culture. In: *The Tinbergen Legacy*. Ed. M. S. Dawkins, T. R. Halliday, and R. Dawkins. London: Chapman & Hall.

Dennett, D. 1991. *Consciousness Explained*. London: Penguin Books.

Dennett, D. 1995. *Darwin's Dangerous Idea: Evolution and the Meanings of Life*. London: Penguin Books.

Desmond, A. 1997. *Huxley: Evolution's High Priest*. London: Michael Joseph.

Devlin, B., Daniels, M., and Roeder, K. 1997. The heritability of IQ. *Nature* 388: 468–71.

Diamond, J. 1991. *The Rise and Fall of the Third Chimpanzee*. London: Vintage.

Dickemann, M. 1979. Female infanticide, reproductive strategies, and social stratification: a preliminary model. In: *Evolutionary Biology and Human Social Behavior: an Anthropological Perspective*. Ed. N. A. Chagnon and W. Irons. Duxbury Press. Pp. 321–67.

Dickinson, A. 1980. *Contemporary Animal Learning Theory*. Cambridge: Cambridge University Press.

Dixson, A. F. 1998. *Primate Sexuality: Comparative Studies of the Prosimians, Monkeys, Apes and Human Beings*. Oxford: Oxford University Press.

Dobzhansky, T. 1937. *Genetics and the Origin of Species*. Columbia University Press: New York.

Dobzhansky, T. 1962. *Mankind Evolving.* New Haven, CT: Yale University Press.

Draper, P. 1989. African marriage systems: perspectives from evolutionary ecology. *Ethology and Sociobiology* 10: 145–69.

Durham, W. H. 1991. *Coevolution: Genes, Culture and Human Diversity.* Stanford, CA: Stanford University Press.

Dwyer, G., Levin, S. A., and Buttel, L. 1990. A simulation model of the population dynamics of myxomatosis. *Ecological Monographs* 60: 423–47.

Eaves, L. J., Eysenck, H. J., and Martin, N. G. 1989. *Genes, Culture and Personality: an Empirical Approach.* San Diego CA: Academic Press.

Edelman, G. M. 1987. *Neural Darwinism: the Theory of Neurological Group Selection.* New York: Basic Books.

Ehrlich, P. R. 2000. *Human Natures. Genes, Cultures, and the Human Prospect.* Washington, D.C.: Island Press.

Endler, J. A. 1986a. *Natural Selection in the Wild.* Princeton, NJ: Princeton University Press.

Endler, J. A. 1986b. The newer synthesis? Some conceptual problems in evolutionary biology. *Oxford Surveys in Evolutionary Biology* 3: 224–43.

Evans, R. I. 1975. *Konrad Lorenz: the Man and his Ideas.* New York: Harcourt Brace.

Falconer, D. S. and Mackay, T. F. C. 1996. *Introduction to Quantitative Genetics. 4th edn.* Harlow: Longman.

Feldman, M. W. and Cavalli-Sforza, L. L. 1976. Cultural and biological evolutionary processes, selection for a trait under complex transmission. *Theoretical Population Biology* 9: 238–59.

Feldman, M. W. and Cavalli-Sforza, L. L. 1989. On the theory of evolution under genetic and cultural transmission with application to the lactose absorption problem. In: *Mathematical Evolutionary Theory*. Ed. M. W. Feldman. Princeton: Princeton University Press. Pp. 145–73.

Feldman, M. W. and Laland, K. N. 1996. Gene–culture coevolutionary theory. *Trends in Ecology and Evolution* 11: 453–7.

Feldman, M. W. and Otto, S. P. 1997. Twin studies, heritability, and intelligence. *Science* 278: 1383–4.

Feldman, M. W. and Zhivotovsky, L. A. 1992. Gene–culture coevolution: towards a general theory of vertical transmission. *Proceedings of the National Academy of Sciences USA* 89: 11935–8.

Feldman, M. W., Aoki, K., and Kumm, J. 1996. Individual versus social learning. *Anthropological Science* 104: 209–32.

Fessler, D. M. T. 2002. Emotions and cost–benefit assessment: the role of shame and self-esteem in risk taking. In: *Bounded Rationality: the Adaptive Toolbox*. Ed. G. Gigerenzer and R. Selten. Cambridge MA: MIT Press. Pp. 191–214.

Fessler, D. 2002. Reproductive immunosuppression and diet: an evolutionary perspective on pregnancy sickness and meat consumption. *Current Anthropology*

Flinn, M. V. 1997. Culture and the evolution of social learning. *Evolution and Human Behavior* 18: 23–67.

Flinn, M. V. and Alexander, R. D. 1982. Culture theory: the developing synthesis from biology. *Human Ecology* 10: 383–400.

Fodor, J. A. 1983. *The Modularity of Mind*. Cambridge MA: MIT Press.

Foley, R. 1996. The adaptive legacy of human evolution: a search for the environment of evolutionary adaptedness. *Evolutionary Anthropology* 4: 194–203.

Forrest, D. W. 1974. *Francis Galton: the Life and Work of a Victorian Genius.* London: Paul Elek.

Futuyma, D. A. 1986. *Evolutionary Biology. 2nd edn.* Sunderland, MA: Sinauer.

Futuyma, D. J. 1998. *Evolutionary Biology. 3rd edn.* Sunderland MA: Sinauer.

Gabora, L. 1997. The origin and evolution of culture and creativity. *Journal of Memetics—Evolutionary Modes of Information Transmission* 1. <http://www.cpm.mmu.ac.uk/jom-emit/1997/vol1/ gabora_l.html>

Galef, B. G. Jr. 1995. Why behaviour patterns that animals learn socially are locally adaptive. *Animal Behaviour* 49: 1325–34.

Galton, F. 1869. *Hereditary Genius.* London: Julian Friedman Publishers.

Galton, F. 1883. *Inquiries into Human Faculty and its Development.* London: Macmillan.

Garcia, J. and Koelling, R. A. 1966. Prolonged relation of cue to consequence in avoidance learning. *Psychonomic Science* 4: 123–4.

Gatherer 1998. Meme pools, World 3, and Averroës's vision of immortality. *Zygon* 33: 203–19.

Gaulin, S. and Schegel, A. 1980. Paternal confidence and paternal investment: a cross-cultural test of a sociobiological hypothesis. *Ethology and Sociobiology* 1: 301–9.

Gigerenzer, G., Todd, P. M., and the ABC Research Group. 1999. *Simple Heuristics That Make Us Smart.* Oxford: Oxford University Press.

Gingerich, P. D. 1983. Rates of evolution: effects of time and temporal scaling. *Science* 222: 159–61.

Goldberger, A. S. 1978. *Models and Methods in the IQ Debate: Part I Revised.* Social Systems Research Institute, University of Wisconsin.

Goodall, J. 1986. *The Chimpanzees of Gombe: Patterns of Behavior.* Cambridge MA: Harvard University Press.

Goodenough, O. R. and Dawkins, R. 1994. The St Jude mind virus. *Nature* 371: 23–4.

Goodenough, W. H. 1999. Outline of a framework for a theory of cultural evolution. *Cross-Cultural Research* 33: 84–107.

Gould, S. J. 1991. *Bully for Brontosaurus. Reflections in Natural History.* New York: Norton & Co.

Gould, S. J. and Vrba, E. 1982. Exaptation: a missing term in the science of form. *Paleobiology* 8: 4–15.

Grafen, A. 1984. Natural selection, kin selection and group selection. In: *Behavioural Ecology: an Evolutionary Approach. 2nd edn.* Ed. J. Krebs and N. B. Davies. Oxford: Blackwell Scientific Publications. Pp. 62–84.

Grant, P. R. and Grant, B. R. 1995. Predicting microevolutionary responses to directional selection on heritable variation. *Evolution* 49: 241–51.

Gruber, H. E. 1974. *Darwin on Man.* New York: Dutton.

Guglielmino, C. R., Viganotti, C., Hewlett, B., and Cavalli-Sforza, L. L. 1995. Cultural variation in Africa: role of mechanism of transmission and adaptation. *Proceedings of the National Academy of Sciences USA* 92: 7585–9.

Haldane, J. B. S. 1955. Population genetics. *New Biology* 18: 34–51.

Haldane, J. B. S. 1956. The argument from animals to man —an examination of its validity for anthropology. *Journal of the Royal Anthropological Institute* 86: 1–14.

Haldane, J. B. S. 1964. A defense of beanbag genetics. *Perspectives in Biology and Medicine* 7: 343–59.

Hamilton, W. 1964. The genetical evolution of social behaviour: I. *Journal of Theoretical Biology* 7: 1–16.

Hamilton, W. 1964. The genetical evolution of social behaviour: II. *Journal of Theoretical Biology* 7: 17–32.

Hamilton, W. D. 1970. Selfish and spiteful behaviour in an evolutionary model. *Nature* 228: 1218–20.

Harpending, H. 1994. Infertility and forager demography. *American Journal of Physical Anthropology* 93: 385–90.

Hartl, D. L. and Clark, A. G. 1989. *Principles of Population Genetics, 2nd edn.* Sunderland, MA: Sinauer.

Hartung, J. 1976. On natural selection and inheritance of wealth. *Current Anthropology* 17: 607–13.

Hartung, J. 1982. Polygyny and the inheritance of wealth. *Current Anthropology* 23: 1–12.

Harvey, P. H. and Pagel, M. D. 1991. *The Comparative Method in Evolutionary Biology.* Oxford: Oxford University Press.

Hawkes, K. 1991. Showing off: tests of another hypothesis about men's foraging goals. *Ethology and Sociobiology* 11: 29–54.

Hawkes, K., O'Connell, J. F., and Blurton Jones, N. G. 1989. Hardworking Hadza grandmothers. In: *Comparative Socioecology: the Behavioural Ecology of Humans and Other Mammals.* Ed. V. Standen and R. A. Foley. Oxford: Blackwell Scientific Publications. Pp. 341–66.

Hawkes, K., O'Connell, J. F., Blurton Jones, N. G., Alvarez, H., and Charnov, E. L. 1998. Grandmothering, menopause, and the evolution of human life histories. *Proceedings of the National Academy of Sciences USA* 95: 1336–9.

Hawkes, K., O'Connell, J. F., Blurton Jones, N. G., Alvarez, H., and Charnov, E. L. 2000. The grandmothering hypothesis and human evolution.' In: *Adaptation and Human Behavior: an Anthropological Perspective.* Ed. L. Cronk, N. Chagnon, and W. Irons. New York: Aldine de Gruyter. Pp. 237–58.

Henrich, J. and Gil-White, F. J. 2001. The evolution of prestige: freely conferred deference as a mechanism for enhancing the benefits of cultural transmission. *Evolution and Human Behavior* 22: 165–96.

Henrich, J., Boyd, R., Bowles, S., Camerer, C., Fehr, E., Gintis, H., and McElreath, R. 2001. In search of Homo economicus: behavioral experiments in 15 small-scale societies. *American Economic Review* 91: 73–7.

Hewlett, B. S. and Cavalli-Sforza, L. L. 1986. Cultural transmission among Aka pygmies. *American Anthropologist* 88: 922–34.

Heyes, C. 2000. Evolutionary psychology in the round. In: *The Evolution of Cognition.* Ed. C. Heyes and L. Huber. Cambridge MA: MIT Press. Pp. 3–22.

Heyes, C. M. and Galef, B. G. 1996. *Social Learning in Animals: the Roots of Culture.* London: Academic Press.

Heyes, C. and Huber, L. (eds.) 2000. *The Evolution of Cognition.* Cambridge MA: MIT Press.

Hill, K. 1988. Macronutrient modifications of optimal foraging theory: an approach using indifference curves applied to some modern foragers. *Human Ecology* 16: 157–97.

Hill, K. and Hurtado, A. M. 1991. The evolution of pre-mature reproductuve senescence and menopause in human females: an evaluation of the 'grandmother' hypothesis. *Human Nature* 2: 313–50.

Hill, K. and Hurtado, A. M. 1996. *Ache Life History: the Ecology and Demography of a Foraging People.* New York: Aldine de Gruyter.

Hill, K. and Hurtado, A. M. 1997. How much does grandma help? In: *Human Nature: a Critical Reader.* Ed. L. Betzig. Oxford: Oxford University Press. Pp. 140–3.

Hinde, R. A. 1974. *Biological Bases of Human Social Behaviour.* New York: McGraw-Hill.

Hinde, R. A. 1982. *Ethology.* Glasgow: Fontana Press.

Hinde, R. A. 1987. *Individuals, Relationships and Culture.* Cambridge: Cambridge University Press.

Hinde, R. A. (ed.) 1991. *The Institution of War.* London: Macmillan.

Hinde, R. A. 1997. War: some psychological causes and consequences. *Interdisciplinary Science Reviews* 22: 229–45.

Hinde, R. A. 1999. *Why Gods Persist: a Scientific Approach to Religion.* London: Routledge.

Hinde, R. A. and Watson, H. E. (eds.) 1995. *War: a Cruel Necessity?* London: Tauris.

Hitler, A. 1943. *Mein Kampf.* London: Pimlico (repr. 1992).

Holden, C. and Mace, R. 1997. Phylogenetic analysis of the evolution of lactose digestion in adults. *Human Biology* 69: 605–28.

Howell, N. 1979. *Demography of the Dobe area !Kung.* New York: Academic Press.

Hrdy, S. B. 1977. *The Langurs of Abu: Female and Male Strategies of Reproduction.* Cambridge MA: Harvard University Press.

Hrdy, S. B. 1981. *The Woman that Never Evolved.* Cambridge MA: Harvard University Press.

Hrdy, S. B. 1999. *Mother Nature: Natural Selection and the Female of the Species.* London: Chatto & Windus.

Hull, D. L. 1982. The naked meme. In: *Learning, Development, and Culture.* Ed. H. C. Plotkin. Chicester: Wiley. Pp. 273–327.

Hull, D. L. 1988. Interactors versus vehicles. In: *The Role of Behaviour in Evolution.* Ed. H. C. Plotkin. Cambridge MA: MIT Press. Pp. 19–50.

Hull, D. 2000. Taking memetics seriously: memetics will be what we make it. In: *Darwinizing Culture: the Status of Memetics as a Science.* Oxford: Oxford University Press. Pp. 43–67.

Huxley, J. S. 1942. *Evolution: the Modern Synthesis.* London: Allen & Unwin.

Huxley, T. H. 1863. *Evidence as to Man's Place in Nature.* London: Williams and Norgate.

Insko, C. A., Gilmore, R., Moehle, D., Lipsitz, A., Drenan, S., and Thibaut, J. W. 1982. Seniority in the generational transition of laboratory groups: the effects of social familiarity and task experience. *Journal of Experimental Social Psychology* 18: 577–80.

Insko, C. A., Gilmore, R., Drenan, S., Lipsitz, A., Moehle, D., and Thibaut, J. 1983. Trade versus expropriation in open groups: a comparison of two types of social power. *Journal of Personality and Social Psychology* 44: 977–99.

Irons, W. G. 1983. Human female reproductive strategies. In: *Social Behavior of Female Vertebrates.* Ed. S. K. Wasser. New York: Academic Press. Pp. 169–213.

Jacobs, R. C. and Campbell, D. T. 1961. The perpetuation of an arbitrary tradition through several generations of laboratory microculture. *Journal of Abnormal and Social Psychology* 62: 649–58.

James, W. 1890. *Principles of Psychology.* New York: Holt.

Jones, S. 1999. *Almost Like a Whale: the Origin of Species Updated.* London: Doubleday.

Kant, I. 1781. *Critique of Pure Reason.* London: Everyman (repr. 1993).

Kaplan, H. 1996. A theory of fertility and parental investment in traditional and modern human societies. *Yearbook of Physical Anthropology* 39: 91–135.

Kaplan, H. and Hill, K. 1985. Hunting ability and reproductive success among male Ache foragers. *Current Anthropology* 26: 131–3.

Kaplan, H. S. and Lancaster, J. B. 2000. The evolutionary economics and psychology of the demographic transition to low fertility. In: *Adaptation and Human Behavior: an Anthropological Perspective.* Ed. L. Cronk, N. Chagnon, and W. Irons. New York: Aldine de Gruyter. Pp. 283–322.

Kaplan, H. S., Lancaster, J. B., Bock, J. A., and Johnson, S. E. 1995. Fertility and fitness among Albuquerque men: a competitive labour market theory. In: *Human Reproductive Decisions.* Ed. R. I. M. Dunbar. London: St Martin's Press. Pp. 99–136.

Kendal, J. R. and Laland, K. N. 2000. Mathematical models for memetics. *Journal of Memetics—Evolutionary Modes of Information Transmission* 3. <http://www.cpm.mmu.ac.uk/jom-emit/2000/vol4/kendal_jr&laland_kn.html>

Ketelaar, T. and Ellis, B. J. 2000. Are evolutionary explanations unfalsifiable? Evolutionary psychology and the lakatosian philosophy of science. *Psychological Inquiry* 11: 1–21.

Kingsolver, J. G., Hoekstra, H. E., Hoekstra, J. M., Berrigan, D., Vignieri, S. N., Hill, C. E., Hoang, A., Gilbert, P., and Beerli, P. 2001. The strength of phenotypic selection in natural populations. *American Naturalist* 157: 245–61

Kitcher P. 1985. *Vaulting Ambition. Sociobiology and the Quest for Human Nature.* Cambridge, MA: MIT Press.

Krebs, J. R. and Davies, N. B. 1997. *Behavioural Ecology: an Evolutionary Approach, 4th edn.* Oxford: Blackwell Science.

Kroeber, A. L. and Kluckholm, C. 1952. Culture: a critical review of concepts and definitions. *Papers of the Peabody Museum of American Archaeology and Ethnology* 47: 1–223.

Kumm, J., Laland, K. N., and Feldman, M. W. 1994. Gene–culture coevolution and sex ratios: the effects of infanticide, sex-selective abortion, and sex-biased parental investment on the evolution of sex ratios. *Theoretical Population Biology* 46: 249–78.

Kuper, A. 1994. *The Chosen Primate.* Cambridge, MA: Harvard University Press.

Kurzban, R. and Leary, M. R. 2001. Evolutionary origins of stigmatization: the functions of social exclusion. *Psychological Bulletin* 127: 187–208.

Lack, D. 1954. *The Natural Regulation of Animal Numbers.* Oxford: Oxford University Press.

Lack, D. 1966. *Population Studies of Birds*. Oxford: Oxford University Press.

Laland, K. N. 1994. Sexual selection with a culturally transmitted mating preference. *Theoretical Population Biology* 45: 1–15.

Laland, K. N. 1999. Exploring the dynamics of social learning with rats. In: *Mammalian Social Learning: Comparative and Ecological Perspectives*. Ed. H. O. Box and K. Gibson. Cambridge: Cambridge University Press. Pp. 174–87.

Laland, K. N. and Odling-Smee, J. 2000. The evolution of the meme. In: *Darwinizing Culture: the Status of Memetics as a Science*. Oxford: Oxford University Press. Pp. 121–41.

Laland, K. N. and Williams, K. 1998. Social transmission of maladaptive information in the guppy. *Behavioral Ecology* 9: 493–99.

Laland, K. N., Kumm, J., and Feldman, M. W. 1995a. Gene–culture coevolutionary theory: a test case. *Current Anthropology* 36: 131–56.

Laland, K. N., Kumm, J., Van Horn, J. D., and Feldman, M. W. 1995b. A gene–culture model of handedness. *Behavior Genetics* 25: 433–45.

Laland, K. N., Odling-Smee, F. J., and Feldman, M. W. 1996a. On the evolutionary consequences of niche construction. *Journal of Evolutionary Biology* 9: 293–316.

Laland, K. N, Richerson, P. J., and Boyd, R. 1996b. Developing a theory of animal social learning. In: *Social Learning in Animals: the Roots of Culture*. Ed. C. M. Heyes and B. G. Galef, Jr. New York: Academic Press. Pp. 129–54.

Laland, K. N., Odling-Smee, J., and Feldman, M. W. 2000. Niche construction, biological evolution, and cultural change. *Behavioral and Brain Sciences* 23: 131–75.

Laland, K. N., Odling-Smee, F. J., and Feldman, M. W. 2001. Cultural niche construction and human evolution. *Journal of Evolutionary Biology* 14: 22–33.

Lande, R. and Arnold, S. J. 1983. The measurement of selection on correlated characters. *Evolution* 37: 1210–26.

Leach, E. 1981. Biology and social science: wedding or rape? *Nature* 291: 267–8.

Leamy, L. 1997. Is developmental stability heritable? *Journal of Evolutionary Biology* 10: 21–9.

Lee, R. B. 1979. *The !Kung San: Men, Women, and Work in a Foraging Society.* Cambridge: Cambridge University Press.

Lehrman, D. S. 1953. A critique of Konrad Lorenz's theory of instinctive behaviour. *Quarterly Review of Biology* 28: 337–63.

Lewontin, R. 1974. *The Genetic Basis of Evolutionary Change.* New York: Columbia University Press.

Lewontin, R. C. 1983a. Gene, organism and environment. In: *Evolution from Molecules to Men.* Ed. D. S. Bendall. Cambridge: Cambridge University Press.

Lewontin, R. C. 1983b. The corpse in the elevator. *The New York Review of Books* 20 January, 29: 34–7.

Lewontin, R. C. 1991. *Biology as Ideology: The Doctrine of DNA.* Toronto: Anasi.

Lewontin, R. 2000. *The Triple Helix: Gene, Organism, and Environment.* Cambridge MA: Harvard University Press.

Li, N., Feldman, M. W., and Li, S. 2000. Cultural transmission in a demographic study of sex ratio at birth in China's future. *Theoretical Population Biology* 58: 161–72.

Lloyd, E. A. 1994. *The Structure and Confirmation of Evolutionary Theory.* Princeton, NJ: Princeton University Press.

346   SENSE AND NONSENSE

Lloyd, E. A. and Feldman, M. W. 2002. Evolutionary psychology: a view from evolutionary biology. *Psychological Enquiry*

Lorenz, K. 1996. *On Aggression*. London: Methuen (repr. in 1996 by Routledge).

Lorenz, K. 1965. *Evolution and Modification of Behavior*. Chicago: University of Chicago Press.

Low, B. S. 2000. Sex, wealth, and fertility: old rules, new environments. In: *Adaptation and Human Behavior: an Anthropological Perspective*. Ed. L. Cronk, N. Chagnon, and W. Irons. New York: Aldine de Gruyter. Pp. 323–44.

Lumsden, C. J. and Wilson, E. O. 1981. *Genes, Mind and Culture: The Coevolutionary Process*. Cambridge, MA: Harvard University Press.

Luttbeg, B., Borgerhoff Mulder, M., and Mangel, M. 2000. To marry again or not: a dynamic model for demographic transition. In: *Adaptation and Human Behavior: an Anthropological Perspective*. Ed. L. Cronk, N. Chagnon, and W. Irons. New York: Aldine de Gruyter. Pp. 345–68.

Lynch, A. (Aaron). 1996. *Thought Contagion: How Belief Spreads through Society*. New York: Basic Books.

Lynch, A. (Alejandro). 1996. The population memetics of birdsong. In: *Ecology and Evolution of Acoustic Communication in Birds*. Ed. D. E. Kroodsma and E. H. Miller. Ithaca: Cornell University Press.

Mace, R. 1996. When to have another baby: a dynamic model of reproductive decision-making and evidence from Gabbra pastoralists. *Ethology and Sociobiology* 17: 263–73.

Mace, R. 2000. An adaptive model of human reproductive rate where wealth is inherited: why people have small families. In: *Adaptation and Human Behavior: an Anthropological Perspective.* Ed. L. Cronk, N. Chagnon and W. Irons. New York: Aldine de Gruyter. Pp. 261–81.

Mackintosh, N. J. 1974. *The Psychology of Animal Learning.* New York: Academic Press.

Mangel, M. and Clark, C. W. 1988. *Dynamic Modeling in Behavioral Ecology.* Princeton, NJ: Princeton University Press.

Markow, T. A. 1995. Evolutionary ecology and developmental instability. *Annual Review of Entomology* 40: 105–20.

Marr, D. 1982. *Vision: a Computational Investigation into the Human Representation and Processing of Visual Information.* San Francisco: Freeman.

Maynard Smith, J. 1964. Group selection and kin selection. *Nature* 201: 1145–7.

Maynard Smith, J. 1975. Survival through suicide. *New Scientist* 28: 496–7.

Maynard Smith, J. 1978. Optimization theory in evolution. *Annual Review of Ecology and Systematics* 9: 31–56.

Maynard Smith, J. and Price, G. 1973. The logic of animal conflict. *Nature* 246: 15–18.

Maynard Smith, J. and Warren, N. 1982. Models of cultural and genetic change. *Evolution* 36: 620–27.

Mayr, E. 1942. *Systematics and the Origin of Species.* New York: Columbia University Press.

Mayr, E. 1963. *Animal Species and Evolution.* Cambridge MA: Harvard University Press.

Miller, G. F. 1997. Mate choice: from sexual cues to cognitive adaptations. In: *Characterising Human Psychological Adaptations. Ciba Foundation Symposium 208*. Chicester: Wiley. Pp. 71–87.

Miller and Todd, 1995. The role of mate choice in bio-computation: sexual selection as a process of search, optimization, and diversification. In: *Evolution and Biocomputation: Computational Models of Evolution*. Ed. W. Banzaf and F. H. Eeckman. Berlin: Springer-Verlag. Pp. 169–204.

Mithen, S. 1996. *The Prehistory of the Mind*. New York: Thames and Hudson.

Møller, A. P. 1990. Fluctuating asymmetry in male sexual ornaments may reliably reveal male quality. *Animal Behaviour* 40: 1185–1187.

Møller, A. P. and Thornhill, R. 1997. A meta-analysis of the heritability of developmental stability. *Journal of Evolutionary Biology* 10: 1–16.

Morgan, C. L. 1896. *Habit and Instinct*. London: Edward Arnold.

Morgan, C. L. 1900. *Animal Behaviour*. London: Edward Arnold.

Morgan, C. L. 1930. *The Animal Mind*. London: Edward Arnold.

Morgan, L. H. 1877. *Ancient Society, or Researches in the Lines of Human Progress from Savagery through Barbarism to Civilization*. New York: Holt.

Morris, D. 1967. *The Naked Ape*. Vintage: London.

Nesse, R. and Williams, G. C. 1995. *Why We Get Sick: the New Science of Darwinian Medicine*. New York: Times Books.

Nisbett, R. E. and Wilson, T. D. 1977. Telling more than we can know: verbal reports on mental processes. *Psychological Review* 84: 231–49.

van Noordwijk, A. J. and De Jong, G. 1986. Acquisition and allocation of resources: their influence on variation in life history tactics. *American Naturalist* 128: 127–42.

Odling-Smee, F. J., Laland, K. N., and Feldman, M. W. 1996. Niche construction. *American Naturalist* 147: 641–8.

Odling-Smee, F. J., Laland, K. N., and Feldman, M. W. 2000. Niche construction and gene–culture co-evolution: an evolutionary basis for the human sciences. *Perspectives in Ethology: Volume 13*. Ed. F. Tonneau and N. S. Thompson. New York: Plenum. Pp. 89–111.

Oldroyd, D. R. 1983. *Darwinian Impacts: an Introduction to the Darwinian Revolution. 2nd edn.* Milton Keynes: Open University Press.

Orians, G. H. 1969. On the evolution of mating systems in birds and mammals. *American Naturalist* 103: 589–603.

Orzack, S. H. and Sober, E. 2001. Adaptation, phylogenetic inertia, and the method of controlled comparisons. In: *Adaptationism and Optimality*. Ed. S. H. Orzack and E. Sober. Cambridge: Cambridge University Press.

Otto, S. P., Christiansen, F. B., and Feldman, M. W. 1995. Genetic and cultural inheritance of continuous traits. *Morrison Institute for Population and Resource Studies* Paper no. 64. Stanford CA: Stanford University Press.

Oyama, S., Gray, R., and Griffiths, P. 2001. *Cycles of Contingency: Developmental Systems and Evolution.* Cambridge MA: MIT Press.

Packer, C., Tatar, M., and Collins, A. 1998. Reproductive cessation in female mammals. *Nature* 392: 807–11.

Pagel, M. 1994. Detecting correlated evolution on phylogenies: a general method for the comparative analysis of discrete characters. *Proceedings of the Royal Society, London Series B* 255: 37–45.

Pagel, M. 1997. Desperately concealing fathers: a theory of parent–infant resemblance. *Animal Behaviour* 53: 973–81.

Palmer, A. R. 2000. Quasireplication and the contract of error: lessons from sex ratios, heritabilities and fluctuating asymmetry. *Annual Review of Ecology and Systematics* 31: 441–80.

Palmer, A. R. and Strobeck, C. 1997. Fluctuating asymmetry and developmental stability: heritability of observed variation vs. heritability of inferred cause. *Journal of Evolutionary Biology* 10: 39–49.

Pawlowski, B. 1999. Loss of oestrus and concealed ovulation in human evolution: the case against the sexual-selection hypothesis. *Current Anthropology* 40: 257–75.

Pawlowski, B., Dunbar, R. I. M., and Lipowicz, A. 2000. Tall men have more reproductive success. *Nature* 403: 156.

Pennington, R. and Harpending, H. 1988. Fitness and fertility among Kalahari !Kung. *American Journal of Physical Anthropology* 77: 303–19.

Pinker, S. 1994. *The Language Instinct.* London: Penguin Books.

Pinker, S. 1997. *How the Mind Works.* London: Penguin Books.

Plomin *et al.* 1993. Genetic change and continuity from fourteen to twenty months: the McArthur longitudinal twin study. *Child Development* 64: 1354–76.

Plotkin, H. C. (ed.) 1982. *Learning, Development, and Culture: Essays in Evolutionary Epistemology.* Chichester: Wiley.

Plotkin, H. 1994. *Darwin Machines and the Nature of Knowledge.* London: Penguin Books.

Plotkin, H. 1997. *Evolution in Mind: an Introduction to Evolutionary Psychology.* London: Penguin Books.

Plotkin, H. 2000. Culture and psychological mechanisms. In: *Darwinizing Culture: the Status of Memetics as a Science.* Oxford: Oxford University Press. Pp. 69–82.

Plotkin, H. C. and Odling-Smee, F. J. 1981. A multi-level model of evolution and its implications for sociobiology. *Behavioral and Brain Sciences* 4: 225–68.

Pocklington, R. and Best, M. L. 1997. Cultural evolution and units of selection in replicating text. *Journal of Theoretical Biology* 188: 79–87.

Popper, K. R. 1979. *Objective Knowledge: an Evolutionary Approach.* Oxford: Clarendon Press.

Price, G. R. 1970. Selection and covariance. *Nature* 277: 520–1.

Proctor, H. C. 1991. Courtship in the water mite *Neumania papillator*: males capitalize on female adaptations for predation. *Animal Behaviour* 42: 589–98.

Proctor, H. C. 1992. Sensory exploitation and the evolution of male mating behaviour: a cladistic test using water mites (Acari: Parasitengona). *Animal Behaviour* 44: 745–52.

Profet, 1988. The evolution of pregnancy sickness as protection to the embryo against Pleistocene teratogens. *Evolutionary Theory* 8: 177–90.

Reader, S. M. and Laland, K. N. 1999. Do animals have memes? *Journal of Memetics—Evolutionary Modes of Information Transmission* 3. <http://www.cpm.mmu.ac.uk/jom-emit/1999/vol3/reader_sm&laland_kn.html>.

Rescorla, R. A. and Wagner, A. R. 1972. A theory of Pavlovian conditioning: variations in the effectiveness of reinforcement and nonreinforcement. In: *Classical Conditioning II: Current Research and Theory.* Ed. A. H. Black and W. F. Prokasy. New York: Appleton. Pp. 64–99.

Reznick, D. N., Shaw, F. H., Rodd, H., and Shaw, R. G. 1997. Evaluation of the rate of evolution in natural populations of guppies (*Poecilia reticulata*). *Science* 275: 1934–7.

Rice, S. H. 1995. A genetical theory of species selection. *Journal of Theoretical Biology* 177: 237–45.

Richards, R. J. 1987. *Darwin and the Emergence of Evolutionary Theories of Mind and Behavior.* Chicago: University of Chicago Press.

Richerson, P. J. and Boyd, R. 1998. The evolution of human ultra-sociality. In: *Indoctrinability, Warfare and Ideology: Evolutionary Perspectives.* Ed. I. Eibl-Eibesfeldt and F. K. Salter. Oxford: Berghahn Books.

Richerson, P. J. and Boyd, R. 2001. The evolution of subjective commitment to groups: A tribal instincts hypothesis. In: *Evolution and the Capacity for Commitment.* Ed. R. M. Nesse. New York: Russell Sage.

Ridley, M. 1997. *Evolution. 2nd edn.* Oxford: Blackwell Scientific Publications.

Rogers, A. R. 1988. Does biology constrain culture? *American Anthropologist* 90: 819–31.

Rogers, A. R. 1990. Evolutionary economics of human reproduction. *Ethology and Sociobiology* 11: 479–95.

Rogers, A. R. 1993. Why menopause? *Evolutionary Ecology* 7: 406–20.

Romanes, G. J. 1882. *Animal Intelligence*. London: Kegan, Paul, Trench & Co.

Rose, H. and Rose, S. (eds.) 2000. *Alas Poor Darwin: Arguments Against Evolutionary Psychology*. London: Jonathan Cape.

Rose, M. R. and Lauder, G. V. 1996. *Adaptation*. San Diego CA: Academic Press.

Rose, S. 1981. Genes and race. Letter. *Nature* 289: 335.

Rose, S., Lewontin, R. C., and Kamin, L. J. 1984. *Not in Our Genes: Biology, Ideology, and Human Nature*. London: Penguin Books.

Royer, C. 1870. *Origine de l'Homme et des Societés*. Paris: Guillaumin.

Ruse, M. 1999. *Mystery of Mysteries: Is Evolution a Social Construction?* Cambridge, MA: Harvard University Press.

Sahlins, M. 1976. *The Use and Abuse of Biology. An Anthropological Critique of Sociobiolgy*. Ann Arbor: University of Michigan Press.

Salzen, E. A. 1996. Introduction to the Routledge edition. In: Lorenz, K. *On Aggression*. Routledge: London.

Schlichting, C. D. and Pigliucci, M. 1998. *Phenotypic Evolution: a Reaction Norm Perspective*. Sunderland MA: Sinauer.

Segerstråle, U. 1986. Colleagues in conflict: an 'in vivo' analysis of the sociobiology controversy. *Biology and Philosophy* 1: 53–87.

Segerstråle, U. 2000. *Defenders of the Truth: the Sociobiology Debate.* Oxford: Oxford University Press.

Sellen, D. W., Borgerhoff Mulder, M., and Sieff, D. F. 2000. Fertility, offspring quality, and wealth in Datoga pastoralists. In: *Adaptation and Human Behavior: an Anthropological Perspective.* Ed. L. Cronk, N. Chagnon, and W. Irons. New York: Aldine de Gruyter. Pp. 91–114.

Shepard, R. N. 1992. The perceptual organization of colors: an adaptation to regularities of the terrestrial world. In: *The Adapted Mind: Evolutionary Psychology and the Generation of Culture.* Ed. J. H. Barkow, L. Cosmides and J. Tooby. Oxford: Oxford University Press. Pp. 495–532.

Sherman, P. W. 1998. The evolution of menopause. *Nature* 392: 759–61.

Shettleworth, S. 2000. Modularity and the evolution of cognition. In: *The Evolution of Cognition.* Ed. C. Heyes and L. Huber. Cambridge MA: MIT Press. Pp. 43–60.

Sieff, D. F. 1990. Explaining biased sex ratios in human populations. *Current Anthropology* 31: 25–48.

Simoons, F. J. 1969. Primary adult lactose intolerance and the milking habit: a problem in biological and cultural interrelations: I. Review of the medical research. *American Journal of Digestive Diseases* 14: 819–36.

Simpson, G. G. 1944. *Tempo and Mode in Evolution.* New York: Columbia University Press.

Sinervo, B. and Basolo, A. L. 1996. Testing adaptation using phenotypic manipulations. In: *Adaptation.* Ed. M. R. Rose and G. V. Lauder. San Diego CA: Academic Press. Pp. 149–85

Smith, E. A. 1985. Inuit foraging groups: some simple models incorporating conflicts of interest, relatedness, and central place sharing. *Ethology and Sociobiology* 6: 27–47.

Smith, E. A. 1992. Human behavioral ecology: II. *Evolutionary Anthropology* 1: 50–5.

Smith, E. A. 1998. Is Tibetan polyandry adaptive? Methodological and metatheoretical analyses. *Human Nature* 9: 225–61.

Smith, E. A. 2000. Three styles in the evolutionary analysis of human behavior. In: *Adaptation and Human Behavior: an Anthropological Perspective.* Ed. L. Cronk, N. Chagnon and W. Irons. New York: Aldine de Gruyter. Pp. 27–46.

Smith, E. A. and Winterhalder, B. (eds). 1992. *Evolutionary Ecology and Human Behavior.* New York: Aldine de Gruyter.

Smith, E. A., Borgerhoff Mulder, M., and Hill, K. 2000. Evolutionary analyses of human behaviour: a commentary on Daly and Wilson. *Animal Behaviour* 60: F21–F26.

Smith, E. A., Borgerhoff Mulder, M., and Hill, K. 2001. Controversies in the evolutionary social sciences: a guide for the perplexed. *Trends in Ecology and Evolution* 16: 128–35.

Smuts, B. B. and Smuts, R. W. 1993. Male aggression and sexual coercion of females in nonhuman primates and other mammals: evidence and theoretical implications. *Advances in the Study of Behavior* 22: 1–63.

Smuts, B. B., Cheney, D. L., Seyfarth, R. M., Wrangham, R. W., and Strusaker, T. T. 1987. *Primate Societies.* Chicago: University of Chicago Press.

Sober, E. and Wilson, D. S. 1998. *Unto Others: the Evolution and Psychology of Unselfish Behavior.* Cambridge MA: Harvard University Press.

Soltis, J., Boyd, R., and Richerson, P. J. 1995. Can group-functional behaviors evolve by cultural group selection? An empirical test. *Current Anthropology* 36: 473–94.

Speel, H. C. 1995. Memetics: On a conceptual framework for cultural evolution. Paper presented at the symposium 'Einstein meets Magritte', Free University of Brussels, June.

Spencer, H. 1855, 1870. *Principles of Psychology 1st edn., 2nd edn.* London: Longman.

Sperber, D. 1996. *Explaining Culture: a Naturalistic Approach.* Oxford: Blackwell.

Sperber, D. 2000. An objection to the memetic approach to culture. In: *Darwinizing Culture: the Status of Memetics as a Science.* Oxford: Oxford University Press. Pp. 163–73.

Stearns, S. C. 1986. Natural selection and fitness, adaptation and constraint. In: *Patterns and Processes in the History of Life: a Report of the Dahlem Workshop.* Ed. D. Jablonski and D. Raup. Berlin: Springer-Verlag.

Stearns, S. 1992. *The Evolution of Life History.* Oxford: Oxford University Press.

Stephens, D. W. and Krebs, J. R. 1986. *Foraging Theory.* Princeton, NJ: Princeton University Press.

Strassmann, B. I. 1981. Sexual selection, paternal care, and concealed ovulation in humans. *Ethology and Sociobiology* 2: 31–40.

Strassmann, B. I. 2000. Polygyny, family structure, and child mortality: a prospective study among the Dogon of Mali. In: *Adaptation and Human Behavior: an*

Anthropological Perspective. Ed. L. Cronk, N. Chagnon, and W. Irons. New York: Aldine de Gruyter. Pp. 49–67.

Sulloway, F. J. 1979. *Freud, Biologist of the Mind: Beyond the Psychoanalytic Legend.* London: Burnett Books.

Symons, D. 1987. If we're all Darwinians, what's the fuss about? In: *Sociobiology and Psychology: Ideas, Issues and Applications.* Ed. C. Crawford, M. Smith, and D. Krebs. Hillsdale NJ: Erlbaum.

Symons, D. 1989. A critique of Darwinian anthropology. *Ethology and Sociobiology* 10: 131–44.

Symons, D. 1990. Adaptiveness and adaptation. *Ethology and Sociobiology* 11: 427–44.

Thompson, J. N. 1998. Rapid evolution as an ecological process. *Trends in Ecology and Evolution* 13: 329–32.

Thorndike, E. L. 1911. *Animal Intelligence.* New York: Macmillan.

Thorpe, W. H. 1979. *The Origins and Rise of Ethology.* London: Heinemann.

Tiger, L. 1969. *Men in Groups.* New York: Random House.

Tiger, L. and Fox, R. 1971. *The Imperial Animal.* New York: Holt, Rinehart, Winston.

Tinbergen, N. 1951. *The Study of Instinct.* Oxford: Oxford University Press.

Tinbergen, N. 1963. On aims and methods of ethology. *Zeitschrift fur Tierpsychologie* 20: 410–33.

Todd, P. M. 2001. Fast and frugal heuristics for environmentally bounded minds. In: *Bounded Rationality: the Adaptive Toolbox.* Ed. G. Gigerenzer and R. Selten. Cambridge MA: MIT Press. Pp. 51–70.

Tooby, J. and Cosmides, L. 1989. Evolutionary psychology and the generation of culture, part I. Theoretical considerations. *Ethology and Sociobiology* 10: 29–49.

Tooby, J. and Cosmides, L. 1990a. The past explains the present: emotional adaptations and the structure of ancestral environments. *Ethology and Sociobiology* 11: 375–424.

Tooby, J. and Cosmides, L. 1990b. On the universality of human nature and the uniqueness of the individual: the role of genetics and adaptation. *Journal of Personality* 58: 17–67.

Trillmich, F. 1996. Parental investment in pinnipeds. *Advances in the Study of Behavior* 25: 533–77.

Trivers, R. L. 1971. The evolution of reciprocal altruism. *Quarterly Review of Biology* 46: 35–57.

Trivers, R. L. 1972. Parental investment and sexual selection. In: *Sexual Selection and the Descent of Man, 1871–1971*. Ed. B. Campbell. Chicago: Aldine. Pp. 136–79.

Trivers, R. L. 1974. Parent–offspring conflict. *American Zoologist* 14: 249–64.

Trivers, R. L. 1985. *Social Evolution*. Menlo Park: Benjamin Cumins.

Trivers, R. L. and Willard, D. E. 1973. Natural selection of parental ability to vary the sex ratio of offspring. *Science* 179: 90–2.

Turke, P. W. 1984. Effects of ovulatory concealment and synchrony on protohominid mating systems and parental roles. *Ethology and Sociobiology* 5: 33–44.

Turke, P. W. 1989. Evolution and the demand for children. *Population and Development Review* 15: 61–90.

Turke, P. W. 1990. Which humans behave adaptively, and why does it matter? *Ethology and Sociobiology* 11: 305–39.

Tylor, E. B. 1865. *Researches into the Early History of Mankind and the Development of Civilization.* London: John Murray.

Tylor, E. B. 1871. *Primitive Culture: Researches into the Development of Mythology, Philosophy, Religion, Art, and Custom, etc.* 2 vols. London: John Murray.

Uyenoyama, M. and Feldman, M. W. 1980. Theories of kin and group selection: a population genetics perspective. *Theoretical Population Biology* 17: 380–414.

Vega-Redondo, F. 1996. *Evolution, Games, and Economic Behaviour.* Oxford: Oxford University Press.

Verner, J. 1964. Evolution of polygamy in the long-billed marsh wren. *Evolution* 18: 252–61.

Verner, J. and Willson, M. F. 1966. The influence of habitats on mating systems of the North American passerine birds. *Ecology* 47: 143–7.

Vining, D. R., Jr. 1986. Social versus reproductive success: the central theoretical problem of human sociobiology. *Behavioral and Brain Sciences* 9: 167–216.

Voland, E. 1998. Evolutionary ecology of human reproduction. *Annual Review of Anthropology* 27: 347–74.

Voland, E. and Stephan, P. 2000. 'The hate that love generated'—sexually selected neglect of one's own offspring in humans. In: *Infanticide by Males and its Implications.* Ed. C. P. van Schaik and C. H. Janson. Cambridge: Cambridge University Press. Pp. 447–65.

Wagner, G. P. (ed.) 2001. *The Character Concept in Evolutionary Biology.* San Diego, CA: Academic Press.

Wallace, A. R. 1869. Geological climates and the origin of species. *Quarterly Review* 126: 359–94.

Washburn, S. L. 1981. Longevity in primates. In: *Aging, Biology and Behavior*. Ed. J. March and J. McGaugh. New York: Academic Press. Pp. 11–29.

Wason, P. 1966. Reasoning. In: *New Horizons in Psychology*. Ed. B. M. Foss. London: Penguin.

Wasson, T. 1987. *Nobel Prize Winners*. New York: Wilson.

Watson, J. B. 1913. Psychology as the behaviorist views it. *Psychological Review* 20: 158–77.

Watson, J. B. 1924. *Behaviorism*. New York: Norton.

Weinrich, J. 1977. Human sociobiology: pair bonding and resource predictability (effects of social class and race). *Behavioral Ecology and Sociobiology* 2: 91–118.

Weiss, K. M. 1981. Evolutionary perspectives on human aging. In: *Other Ways of Growing Old*. Ed. P. Amoss and S. Harrell. Stanford: Stanford University Press. Pp. 25–58.

Westneat, D. F. and Sargent, R. C. 1996. Sex and parenting: the effects of sexual conflict and parentage on parental strategies. *Trends in Ecology and Evolution* 11: 87–91.

Whiten, A., Goodall, J., McGrew, W. C., Nishida, T., Reynolds, V., Sugiyama, Y., Tutin, C. E. G., Wrangham, R. W., and Boesch, C. 1999. Cultures in chimpanzees. *Nature* 399: 682–5.

Wilkinson, G. S. 1984. Reciprocal food sharing in the vampire bat. *Nature* 308: 181–4.

Williams, G. C. 1957. Pleiotropy, natural selection, and the evolution of senescence. *Evolution* 11: 398–411.

Williams, G. C. 1966. *Adaptation and Natural Selection: a Critique of Some Current Evolutionary Thought*. Princeton: Princeton University Press (repr. 1996).

Williams, G. C. 1992. *Natural Selection: Domains, Levels, and Challenges.* Oxford: Oxford University Press.

Wilson, D. S. 1999. Flying over uncharted territory [review of *The Meme Machine* by Susan Blackmore]. *Science* 285: 206.

Wilson, E. O. 1975a. *Sociobiology: the New Synthesis.* Cambridge MA: Harvard University Press.

Wilson, E. O. 1975b. Human decency is animal. *The New York Times Magazine* 12 October, 38–50.

Wilson, E. O. 1978 *On Human Nature.* Cambridge, MA: Harvard University Press.

Wilson, E. O. 1994. *Naturalist.* Washington, DC: Island Press.

Wilson, E. O. 2000. Sociobiology at the end of the century. In: *Sociobiology: the New Synthesis. 25th Anniversary edn.* Cambridge, MA: Harvard University Press.

Winterhalder, B. and Smith, E. A. 2000. Analyzing adaptive strategies: human behavioral ecology at twenty-five. *Evolutionary Anthropology* 9: 51–72.

Wright, R. 1994. *The Moral Animal: Why We Are The Way We Are (The New Science of Evolutionary Psychology).* London: Abacus.

Wynne-Edwards, V. 1962. *Animal Dispersion in Relation to Social Behaviour.* Edinburgh: Oliver and Boyd Ltd.

Zahavi, A. 1975. Mate selection—selection for a handicap. *Journal of Theoretical Biology* 53: 205–14.

Index

Q - What proximate mechanisms favor the accomplishment of ultimate mechanisms/aims?

1. Self-preservation.
   a. greater labor among females than males? (Or does it adjust?)
   b. Drive greatest during years of peak reproduction & years of greatest need on part of offspring for parental health and investment.

Q - for humans, aren't there class/economic constraints that make these ages, levels, etc. different within a given society?

Q - of what use is human ability to cross normative boundaries.